G. Shuckburgh

Observations Made in Savoy,

in Order to Ascertain the Height of Mountains by means of the Barometer;

Being an Examination of Mr. De Luc's Rules Delivered in His Recherches

Sur les Modifications de l'Atmosphere. By Sir George Shuckburgh, Bart. F.

R. S

G. Shuckburgh

Observations Made in Savoy,
in Order to Ascertain the Height of Mountains by means of the Barometer; Being an Examination of Mr. De Luc's Rules Delivered in His Recherches Sur les Modifications de l'Atmosphere. By Sir George Shuckburgh, Bart. F. R. S

ISBN/EAN: 9783337382117

Printed in Europe, USA, Canada, Australia, Japan

Cover: Foto ©berggeist007 / pixelio.de

More available books at **www.hansebooks.com**

XXIX. *Obfervations made in* Savoy, *in order to afcertain the height of Mountains by means of the Barometer; being an Examination of Mr.* De Luc's *Rules, delivered in his* Recherches fur les Modifications de l'Atmofphere. *By Sir* George Shuckburgh, *Bart. F. R. S.*

Read May 8 and 15, 1777. IN the courfe of my tour into Italy in the years 1775 and 1776, I made fome ftay at Geneva; which being in the neighbourhood of the Alps, and on that account a convenient home, induced me to make fome obfervations upon thofe mountains, which have been defervedly objects of attention to the moft incurious traveller. I was particularly defirous of verifying the experiments with the barometer, in taking heights of different fituations; a method that has been long known to the ingenious, though but rarely practifed, and capable of but little precifion till within thefe few years; and perhaps at prefent not fo generally' known as the convenience and utility of the method feems to require. I had provided myfelf with a confiderable collection of inftruments, or a kind of portable philofophical cabinet, which I had had

Vol. LXVII.   X x x   made

made exprefsly in London and Paris, in order to make ſuch experiments as might preſent themſelves to me *en courrant*; and which, either from want of acquaintance with the ſubject, want of time, or want of money, become rarely the object of travellers; but remain wholly unknown till princely munificence and philoſophic zeal (of which we have a recent inſtance) unite in producing them to the world. After the very celebrated and ingenious labours · of Mr. DE LUC, farther inveſtigation of the ſubject of barometrical meaſurement might ſeem unneceſſary, if not invidious; but, furniſhed as I was with an apparatus every way ſufficient for the inquiry, finding myſelf in the country which had been the ſcene of his operations, and poſſeſſing ſome ſhare of his own zeal, I could not but gratify the curioſity I had to verify and repeat his experiments: if therefore in the purſuit of this inquiry I ſhould be led to a concluſion ſomething different from the reſult of· his own obſervations, I am convinced that this diſtinguiſhed obſerver, of whoſe candour and talents I have an equal opinion, will impute it wholly to a love for truth; as with me the precept applies as ſtrongly to the philoſopher as to the hiſtorian, *Ne quid falſi audeat, ne quid veri non audeat dicere.*

But

But to proceed. The instruments I made use of in these operations were, two of RAMSDEN's barometers [a]; three or four thermometers detached from the barometers, whose boiling and freezing points I had examined myself; an equatorial instrument, the circles of which were about seven inches diameter, made by RAMSDEN; a fifty-feet steel measuring chain; and three three-feet rods, two of deal and one of brass, in order to examine and correct the chain, these latter made by BARADELLE at Paris. Besides these I took with me a little bell-tent, which I found of great use, as it defended me from the wind and sun; and I may remark, that the observations of the uppermost barometer were made in the tent.

My first series of observations I proposed to be on Mont Saleve [b], one of the Alps, situated about two

(a) It may not be improper to remark, that the specific gravity of the quick-silver of these barometers with 68° of heat was 13,61; the diameter of the bore of the tube 0,20 inch; and that of the reservoir 1,5 inch.

(b) Mont Saleve extends near nine miles in length; is not quite 3300 feet in height above the Lake. That side of it which is next Geneva is for the most part a barren rock, the north-east end of it being almost a perpendicular precipice; the other side of the mountain is less rude, of a more gentle accli-vity, covered with trees, shrubs, and herbage, as is also the top, where is some of the finest pasture in the world. It is inhabited only by a few shepherds, who pass the summer months here with their cattle, in little miserable huts or barns: the remaining part of the year, viz. for about four or five months, it is covered with snow. This mountain contains chiefly a calcareous stone; and there is reason to believe that there is an iron ore in it, at least in some parts of it, as a piece Mr. DE LUC, the brother, picked up near the south-west end, I found, sensibly affected the magnet.

leagues

leagues fouth of Geneva, and precifely on the fame point where Mr. DE LUC had made his higheft or fifteenth fta- tion: this fpot I learnt from his brother, whofe civilities, both then and fince, I fhall frequently have occafion to remember and mention.

The place where I meafured my bafe was in a field near the villages of Archamp and Neidens, not quite three miles in a horizontal line from the top of the rock whofe height was to be determined (fee the chart that accompanies this account). At the end of the bafe A I intended to place one of my barometers; and the other at the top of the rock, called the Pitton, at c; and with the above inftruments meafure the triangle ABC. The angles were taken both on the horary circle, which was brought parallel to the horizon, and alfo on the azimuth circle of the equatorial inftrument; this made it, as it were, two different inftruments independant of each other. The angles were moreover doubled, tripled, and quadrupled, on each arch; by this means the error of the center or axis of the inftrument vanifhed; the poffi- ble error in the divifions, in the reading off, and in the coincidence of the wires in the telefcope (which magni- fied forty times) with the fignals placed at each angle of the triangle, was leffened in proportion to the number of times the obfervation was repeated; and finally the

mean

mean of all was taken. The fame was done with each angle at A, B, and c, horizontal as well as vertical, *viz.* the elevation of c above A and B was taken; and alfo the depreffion of A and B below c. The advantage of this method was, that the error of the line of collimation, the effect of refraction, and of the curvature of the earth's furface, all became equal and contrary; by thefe means the little errors were diminifhed, and great errors abfolutely avoided[c]. I fhall, however, beg leave to fet down the operation at length refpecting this one triangle, in order to fhew the precifion that may be expected from fuch a geometrical procefs; to remove the fcruples of thofe gentlemen who fufpect that accuracy is only to be obtained by large quadrants; and laftly, to do juftice and fatisfaction to the celebrated artift who invented and made this valuable inftrument.

(c) I muft acknowledge here, that the attraction of the mountain creeps into the account uncorrected for, but only half of this quantity influences the mean refult, as at the top it was nothing; and at the bottom of the mountain it could not exceed 10″ in the direction AC, as I find from a rough computation, the half of which = 5″ would give only four inches for the correction.

Deter-

Determination of the Base.

	Ch.	Ft.	In.	Temper.
Length of the base AB (see the Chart) by the chain, first time, — — — —	55	10	0	71°
Ditto, second time, — — — —	55	9	9¾	76
The mean, — —	55	9	10.87	73½

	Ft.	In.	
By frequent previous observations I determined (4) the length of the chain by comparison with the brass standard rod reduced to 60° of heat, —	50	0	60
Correction for 13½° of heat from expansion, —	+0	0	05
Diameter of the pins or arrows, one of which was used at each chain, and in such manner, that this correction became always + — — —	+0	0	16
Correct length of the chain as it was used in measuring the base, — — — —	50	0	21
Multiply by the number of entire chains in the base, —		55	
	2750	11	55
Add the parts of a chain, — —	+ 9	10	87
True length of the base, as it was measured, —	2760	10	42
Correction for the defect of level, taken with an instrument made on purpose, each time the chain was placed,	— 0	76	(*)
The true horizontal distance between A and B becomes,	2760	9	66

Deter-

(d) It may be required, to what precision I could determine the length of my chain? I think certainly to within $\frac{1}{100}$ of an inch, or $\frac{1}{6000}$ of the whole length. The common GUNTER's chain of the shops is always subject to spring and stretch considerably; mine was made of hardened steel, on purpose to avoid this defect. It however still preserved some degree of elasticity, for when pulled with a force of about ten pounds, it seemed = 0,12 inch longer than when laid gently on the floor without being stretched at all: the assumed length of the chain was such as seemed to me probable from a moderate tension in

common

Determination of the angles by the equatorial.

	On the azimuth circle.	On the equat. circle, the horary being converted into gradual divisions.
∠ A by the 1st observation —	58° 27′ 30″	58° 28′ 30″
2d, — —	— 29 0	— 27 30
3d, — —	— 28 30	— 29 15
4th, — —	— 30 15	— 29 15
∠ taken four times over on the arch, —	233 54 15	233 54 30
The mean, —	58 28 49	58 28 37¼

Lastly, the mean of all from the two circles
≈ 58° 28′ 43¼″ = ∠ at A.

∠ B by the 1st observation, —	111 54 45	111 53 0
2d, —. —	— 51 30	— 52 30
3d, — —	— 50 30	— 50 45
∠ taken three times over on the arch, —	335 36 45	335 36 15
Mean, — —	111 52 15	111 52 5.

Mean of all from the two circles = 111° 52′ 10″
= ∠ at B.

common using it. It may perhaps not be out of place to remark here, that the rods with which the chain was examined, agreed exactly with the scales of the barometers; at least the difference in nine inches, taken in different parts of the scale, did not appear to exceed $\frac{1}{100}$ of an inch.

(*) The precaution in taking the inclination of the chain every time; if the base be nearly a plain, as is the case in many meadows, seems to be unnecessary; for this same correction, deduced from the inclination of the base observed at A and B, comes out —0,99 inch, only 0,23 inch different, a quantity wholly inconsiderable.

∠ C

			On the azimuth circle.	On the equat. circle.
			° ′ ″	° ′ ″
∠ c by the 1st observation,	—		9 39 0 —	9 38 30
2d,	—	—	– 39 0 —	– 38 15
3d,	—	—	– 38 45 —	– 39 45
∠ taken four times over on the arch,	—		38 35 45 —	38 34 45
Mean,	—	—	9 38 56½ —	9 38 41½

Mean of the two circles, = 9° 38′ 48¾″ = ∠ at c.

	By actual observation.		Angles finally corrected.
	° ′ ″		° ′ ″
∠ at A, — —	58 28 43¼	These angles corrected by adding 6″ to each (the sum of their errors, or defect from 180° being —18″) become,	58 28 49¼
∠ at B, — —	111 52 10		111 52 16
∠ at C, — —	9 38 48¼		9 38 54¼
Sum of the three angles =	179 59 42	Sum,	180 0 0
Taken from	180 0 0		
Leaves the difference = } fum of the errors, }	— 18		

It is highly curious and satisfactory to see the amazing correspondency of these observations, made with an instrument of only 3½ inches radius, whereon an angle of one minute is about equal $\frac{1}{1300}$ inch; and I think we may fairly conclude, that the corrected mean result of these observations is true to within 6″ or 8″[f]; which, as

[f] I may have a future occasion to speak of the accuracy of this instrument for astronomical purposes; but I cannot omit this opportunity of mentioning one, viz. in taking the latitude of the city of Amiens in Picardy, where I had thirteen observations by the stars and Sun, the mean of which differed 25″ from the extremes, and only 3″ from the result of Mr. CASSINI's observations, made, I believe, with a nine-feet zenith sector, as related in *La Meridienne de Paris verifiée.*   5

may

may be proved hereafter, would occafion an error of only three feet in the diftance of the mountains, and feven inches in the height. I proceed next to the vertical angles.

Determination of the inclination of the fides AC, BC, *and* AB, *with the horizon; the height of the eye at the inftrument being four feet above the ground.*

Altitude from below at A.			Depreffion from above at C.		
	o ′ ″			o ′ ″	
Inclination of AC, —	10 33 2		Correct for the fignal, —	10 29 18	
Correction for the part of the fignal which was obferved, }	— 1 38		——— for the line of collimation, — }	+ 16 + 59	
Correction for the line of collimation, — }	— 0 59		——— for refraction,	+ 27	
Correct for the refraction, —	0 27		True depreffion of A from C,	10 31 0	
True Altitude of C from A,	10 29 58		Arch intercepted between, or curvature, — }	— 2 30	
			True altitude of C from A deduced from the obfervation at C, }	10 28 30	

Mean corrected altitude of C from A = 10° 29′ 14″ (g).

(g) If the computation were to be made from either of the obfervations taken feparately, the difference would amount to only three feet in the height of C; and this may either be in the correction of the line of collimation, the effect of refraction, or in miftaking the part of the fignal that was obferved: for, whilft I was gone to the top of the mountain, fome peafants poffeffed themfelves of the handkerchiefs I had fixed to the fignals below in order to have a confpicuous and determined point.

Altitude from below at B.			
	o	´	´´
Inclination of BC, —	11	20	26
Correct for the part of the ſignal obſerved, — }		— 1	38
Error of collimation,		— 0	59
Correct for refraction,		— 0	26
True altitude of c from B,	11	17	23

Depreſſion from above at C.			
	o	´	´´
	11	19	47
Correct for the ſignal, —		—	59
Error of collimation, —		+	59
Effect of refraction, —		+	26
True depreſſion of B from c,	11	20	18
Arch intercepted, or curvature, — }		— 2	18
True altitude of c from B, deduced from the obſervation at c, — }	11	18	0

Mean of the two, or corrected altitude of c from B
$= 11° 17' 41\frac{1}{2}''$.

Altitude at A.			
	o	´	´´
∠ of inclination of AB the baſe, — }	0	27	0
Error of the line of collimation, — }		— 0	59
Correct altitude of B from A,	0	26	1

Depreſſion at B.			
	o	´	´´
Error of collimation,	0	27	4
	+ 0		59
Correct depreſſion of A from B; — }	0	28	3
Arch intercepted, —		— 0	27
Altitude of B from A deduced from the obſervation at B, — }	0	27	36

Mean of the two, or corrected altitude of B from A
$= 0° 26' 49''$[b].

(b) It ſhould ſeem from theſe two obſervations, that the error of the line of collimation had been aſſumed too great; it has however, as I have before obſerved, nothing to do with the mean reſult: and this is, perhaps, one of the beſt means of diſcovering the error of collimation, and the very method Mr. DE LUC uſed, to adjuſt his level, though, as I have been informed by his brother, without taking into the account the effect of curvature, which, if his horizontal marks were 2000 feet diſtant from each other, would amount to 20'', and the error to half that quantity.

5   I have

I have thus, in a manner rather prolix, given a detail of the methods used to ascertain the quantity of the different angles. It may be of use on a like occasion, and will at least serve to determine within what limits the error of the final result may be expected to lye, as on the precision of the geometrical operations all the comparisons of the barometrical ones depend. This process once mentioned will exempt me and the reader from the trouble a second time, when he is informed, that the same fidelity and pains were employed (where the circumstances would admit) in all the trigonometrical observations, of which the annexed chart is a summary. I proceed now to the determination of the sides, the computations of which are too well known to enter into this paper.

	Feet.
Side AB	2760.8
AC	15286.4
BC	14041.7

	Feet.
These with the angles give for the height of c above A, —	2835.07
The height of c above B, — — —	2806.27
The height of B above A, — — —	22.18
These two added give the height of c above A, deduced from the observation at B, — — —	2828.45
But the height by actual observation at A was, — —	2835.07
Then the mean of the two, — — —	2831.76

which is probably within three or four feet of the truth, or about one foot in a thousand.

Having

Having thus the perpendicular height, as I think, very accurately afcertained, it remained for me to take the altitude of the barometer at each ftation A and c, and if poffible with equal precifion. Thefe obfervations it would be too tedious to fet down at length. I fhall, however, premife the following particulars. Every obfervation of the barometer was triple; that is, the height was read off three different times, and the mean taken; but from once reading only I could be fure of the height to $\frac{4}{1000}$ of an inch, exclufive of the error of the divifions, which in fome places might amount to that quantity; this the nonius would itfelf difcover and even correct by eftimation. At every feries of obfervations the float at the bottom was readjufted, fo that I could conftantly be fure of an alteration of the weight of the atmofphere expreffed by 0.002 inch of quickfilver, if not of half that quantity. Finally, the difference of the two barometers [i] was conftantly taken, after being left three-quarters

ters

(i) It may be concluded, that this difference fhould be conftant, and always the fame; but, from what caufe I know not, it did not appear fo to me. In my journal for the weather for 1775, I find the following note: from a mean of feventeen obfervations between Auguft 12th and Sept. 1. viz. before, at, and after, my expedition to Mont Saleve and the Mole, I find the difference between my two barometers =,0042 inch, N° 1. ftanding the higheft; in thefe comparifons, however, the extremes fometimes differed from the mean =,006. And in my paffage over Mont Cenis, Dec. 1. barometer N° 1. ftood lower than N° 2.

by

ters of an hour or more in the fame place, to acquire the true temperature of the air, and this before and after every expedition. The fractional parts of a degree on both the attached and detached thermometers were noted only by eftimation, but written down to 10ths, being more convenient in the computation; for I may remark, that one-third of a degree on the attached thermometer is equal to about $\frac{1}{1000}$ inch on the barometer; this attention, therefore, to the fub-divifions of the degrees became neceffary. I conclude, laftly, with prefuming, that the weight [k] of any column of air may be meafured with thefe barometers to ,008 inch, though all the errors fhould lye the fame way.

Leaving Geneva about half paft fix in the morning, Auguft 20th, I arrived at the place A of my bafe a little before eight; near to which there happened to be a fhepherd's houfe, in which I left one of my barometers (N° 1.) with a fervant, to examine and obferve it every five or ten minutes for near nine hours fucceffively,

by —,013 inch : it is difficult to account for this. May 10th, 1776, at Rome, N° 1. ftood loweft by —,001. June 12th, at Naples, N° 1. ftood loweft by —,008. Sept. 10th, in London, N° 1. ftood higheft by +,006. Thefe apparent variations may poffibly arife from fome alteration in the frame-work of the barometers through moifture, &c.

(k) I muft not be underftood to mean, that the length of any column of air may be meafured to an equal accuracy, even though our theory fhould be perfect: this will be the fubject of inquiry in its proper place.

untill

until I returned; the windows and doors of the room,
in which the inſtrument was placed, being left open, by
which means there was a free communication with the
outward air, and the barometer not expoſed to the Sun.
The detached thermometer was hung on the window
towards the north-eaſt, where there was neither direct
nor reflected heat from the Sun [1]. The two barometers

(1) I have thought proper to mention this, as it is almoſt the only circumſtance wherein my method of obſerving differed from Mr. DE LUC's, whoſe
thermometers (if I miſtake not) were hung always in the Sun, and probably
for this reaſon, becauſe the column of the atmoſphere between the two barometers, whoſe mean heat is to be determined, is (if the Sun ſhine) all expoſed
to the Sun. I have, however, always preferred hanging them in the ſhade,
and I give the following reaſons: all ſpurious and local heat from reflection is
more eaſily avoided; no concentrated and falſe heat is acquired by the mounting,
and thence communicated to the tube, even though the ball ſhould be inſulated;
and, finally, becauſe I ſuſpect the real temperature of the atmoſphere in the
Sun and in the ſhade to be the ſame, or at leaſt inſenſibly different. This may
be thought to be advancing too much; but, to be ſatisfied of the poſition, I made
no leſs than four-ſcore obſervations with four different thermometers of very
different mounting, hung alternately expoſed to the Sun's rays, and ſcreened
from them by the ſhade of a tree, in an open plain at ſome diſtance from the
town of Geneva. The reſult was, that my beſt thermometer, with the ball
inſulated, differed only 2° in the different ſituations; the others, more or leſs,
as they were more or leſs connected with the frames in which they hung. One
of them, incloſed in a glaſs tube, roſe 12° higher than the true temperature,
which was 77°. It ſhould ſeem then, that the variety in the mounting occa-
ſioned this difference; and this effect of the materials, of which the inſtrument is made, cannot be wholly avoided, as the glaſs itſelf, which conſtitutes
the ball of the thermometer, will acquire and contain more or leſs, in proportion
to its thickneſs and opacity. If a thermometer were perfect, it would reflect
all the rays that it receives. More might be added to corroborate this idea,
but it would ſwell this note to an unwarrantable length.

were

were here compared; and at a quarter after nine, beginning my walk, I arrived, not without fome fatigue, at
the top of the mountain about noon. The view from
thence was incredibly beautiful. Every object, that from
Geneva was ftriking, from thence appeared with an ad-
ditional effect. The mountains feemed higher and
nearer; the plain appeared a more perfect level, the fmall
inequalities from this height becoming infenfible; and
a larger portion of the lake prefented itfelf: behind me
an innumerable collection of naked points and precipices,
all new objects, that from below are hid by the mountain,
afforded frefh and moft aftonifhing ideas of this very
fingular part of the creation. The clouds however (for
it was a little hazy) unfortunately prevented my feeing
Mont Blanc and the Glacieres, which were ftill farther
behind. Some of the clouds were below me, and very
near; exhibiting to me, at that time, a very fingular phe-
nomenon of the thunder grumbling under my feet. I
was occupied here between four and five hours with different obfervations. The barometrical ones I am now
going to relate; and I fhall at the fame time give the
computations of them according to Mr. DE LUC's method, or rather according to Dr. HORSLEY's reduction of
it to the fcales and meafures of this country *(vide*
Philof. Tranf. vol. LXIV.) with this difference, that I
have

have reckoned the equation for the expanfion of quick-filver =,00323 inch for every degree of FAHRENHEIT's thermometer in a column of 30 inches, inftead of ,00312 which Mr. DE LUC ufed; the former I had collected from fome of my own experiments made at Oxford in the beginning of the year 1773: this difference will not, however, occafion an alteration in the refult of any one of my obfervations of more than five inches, and may therefore be confidered as of no account. Of the real value of this correction I fhall fpeak more hereafter.

The barometer was fet up on the mountain at one o'clock, and left an hour and a quarter to acquire the temperature of the tent in which it was placed, before the firft regular feries of obfervation was taken. The fucceeding obfervations were made at intervals of near an hour each. I have ventured to fet down the height of the barometer to ,0001 inch; but this is only the mean from three or four readings off. It feems that the heat of the tent was confiderably greater than that of the external air; this, however, can only influence the ex-panfion of the quickfilver, fhewn by the attached ther-mometer, and not the preffure of the atmofphere. Laftly, the true difference in the height of the refervoirs of the two barometers, by comparifon with A and c, was found equal 2831.3 feet geometrically.

Com-

Comparifon of the firſt ſeries.

Obſervations at the top of the mountain at c.

	Barom. N° 2. above at c. In. Pts.	Therm. attachod.	Therm. detached.
	25.7120	78.0	65.0
Correct for the diff. of the 2 attached therm. 5°.9,	— 162		
Barometer at the top,	25.6958	Log.	4098621
—————— below,	28.3951	Log.	4532434

Difference, or fall of the quickfilver, } 2.6993 Diff. of Log. 433.813 { the height in Engliſh fathoms,

Correct for 29°.7 of heat, — — + 28.728

Correct height in fathom, — — 462.541 × 6

Height in Englifh feet by the barometer, — 2775.246
Height by the trig. method, — — 2831.3

Difference, or error of the barometer $\frac{181}{1000}$, — 56.1

Obſervations below at A.

	Barom. N° 1. below at A. In. Pts.	Therm. attached.	Therm. detached.
	28.3990	72.1	73.9
Correct for the diff. of the barometer,	— 39		65.0 heat at c.

69.4 mean heat of the air.

'28.3951 39.7 { ſtand. temp. accordin to Dr. HORSLEY.

+ 29.7 difference.

A detached thermometer in the tent ſtood at 72°.

Comparison of the second Series.

Observation at the top of the mount at c.

	Barom. N° 2. above at c. In. Pts.	Therm. attached.	Therm. detached.
	27.7025	73.4°	64.0°
Correct for the Diff. of the two attached therm.	— 50		
Barometer at the top,	25.6975	Log. 4098908	
———— below, —	28.3901	Log. 4531669	

Difference, or fall of the quickfilver, — } 2.6926 Diff. of Log. 432.751 { approx. height in English fathoms.

Correct for 28°.8 of heat, — — + 27.787

Corrected height in fathoms, — — 460.538 × 6

Height in feet by the barometer — 2763.228
———— by the trig. method, — 2831.3

Difference, or error of the barometer $\frac{248}{1000}$, — 68.1

Observation below at A.

	Barom. N° 1. below at A. In. Pts.	Therm. attached.	Therm. detached.
	28.3940	71.6°	73.0°
Correct for the diff. of barometer,	— 39		64.0 heat at c.
	28.3901		68.5 mean heat.
			39.7 standard temperature.
			+ 28.8 difference.

A detached thermometer in the tent stood at 69°.

During these observations the wind was S.W.; the weather hazy, accompanied with a little thunder.

Com-

Comparison of the third series.

Observations at the top near c.

	Barom. N° 2. above at c. In. Pts.	Therm. attached.	Therm. detached.
	25.6900	69.7	62.0
Correct for the diff. of the 2 attached therm. 1°.4, }	+ 38		
Barometer at the top,	25.6938	Log. 4098283	
—————— below, —	28.3896	Log. 4531593	
Difference, or fall of the quicksilver, — }	2.6958	Diff. of Log. 433.310	{ Approx. height in fathoms.
Correct for 27°.5 of heat,	— —	+ 26.582	
Correct height in fathoms,	— —	459.892	
		× 6	
Height in feet by the barometer,	—	2759.352	
—————— by the trig. method,	— —	2831.3	
Difference, or error of the barometer $\frac{713}{1000}$,		— 71.9	

Observations below near to A.

	Barom. N° 1. below at A. In. Pts.	Therm. attached.	Therm. detached.
	28.3935	71.1.	72.5
Correct for the diff. of barometer, }	— 39		62.0 heat at c.
			67.2 mean heat.
	28.3896		39.7 standard temperature.
			+ 27.5 difference.

A detached thermometer in the tent stood at 65°.

These

These observations then seem to prove that the barometrical rules were a little defective as to the true ratio between the gravities of air and quickfilver, *viz.* in the value of an inch of quickfilver in the torricellian tube, expressed in inches of the atmosphere with a given temperature. The first comparison gives for this error in defect −19.8 feet in every 1000 feet; the second, 24.0 feet; and the last, 25.4 feet: the mean of the three is 23.1 feet; and by so much we may conclude that these rules, in greater heights also, will give the difference of elevation too little, *viz.* by $\frac{1}{43}$ nearly [m]. But it will be fair to make the experiment.

(m) Left any suspicion should arise of a disagreement between the actual measures taken by Mr. DE LUC and myself, I may observe, that the mean result of three observations, which I made independently of each other on the height of the Pitton or point c above the lake of Geneva, agree with the mean result of Mr. DE LUC's operation from the levelling and the quadrant, to less than twelve inches; a greater correspondency than which cannot be expected: and this was the true reason why I chose the same spot he had pitched upon. " *Le rocher* " *isolé, qui domine toute la montagne.*" As a further confirmation, I compared his standard steel rod of twelve Paris inches, which his brother obligingly furnished me with, with my brass one, and found twelve inches on Mr. DE LUC's rule was on my rule, with 71° of heat, — — 12.784 Eng. inches.

Correction for the difference of expansion between brass and steel with 16° of heat, — — } + 07

Length of Mr. DE LUC's French foot with 55°, — 12.7847
True length of the French foot *(vide* Phil. Tranf.) 12.7890

Error or difference from the true Paris foot — −,0043 $= \frac{1}{1000}$ nearly.

The Mole is a convenient, infulated mountain, fituated about eighteen miles eaft of Geneva, and rifing near five thoufand feet above the lake, much higher than any body, that I know of, has ever made thefe experiments at, with the required precifion. On this fummit I determined to confirm or correct my difcovery, and communicated my intentions to Mr. DE SAUSSURE, a very ingenious gentleman of this place, and well fkilled in various parts of natural and experimental philofophy, who gave me all the information neceffary, and obligingly promifed to accompany me, as did alfo Mr. TREMBLEY, affiftant to Mr. MALLET, well known in the aftronomical world. This expedition was undertaken in the latter end of Auguft and beginning of September. I fhall here beg leave to fet the reader down at the bottom of the mountain, and flatter myfelf he will accompany me to the top. It was about five in the afternoon when we left St. Joire, a wretched little village at the foot of the mountain to the eaft, and where we had dined in a moft miferable *auberge*, preparing to afcend the fummit on foot, being feven or eight in company, including guides and fervants, who carried my inftruments, provifions, &c.; the former confifting of the equatorial, the barometer, different thermometers, electrical balls, an hygrometer, and a dipping-needle; together with another barometer

of

of Mr. DE LUC's conftruction, a variation-needle, a level belonging to Mr. DE SAUSSURE, and a tent. Thus accoutered we proceeded up an afcent, not however very fteep, for three hours and a half without intermiffion, the path leading in a fpiral kind of direction, very rugged and full of loofe pieces of rock that are brought down with the melting fnows, paffing through romantic woods of fine firs and other trees, interfperfed here and there with a thin foil of excellent pafture. Before we arrived at the hut, where we were to fleep (for our intention was to lay upon the mountain that night, in order to have the more time the next morning for our operations) having walked on a little too far before, we loft fight of our guides. We called feveral times, but were never anfwered:—the night was now coming on; a kind of fog appeared, with fmall rain; our fituation became fomewhat embarraffing. We called again, but were anfwered by nothing but an echo, the place being a moft profound folitude. We began now to confider ourfelves as loft. Mr. DE SAUSSURE, though he had been feven or eight times before upon the mountain, found himfelf in doubt concerning the way; but after a fhort dilemma thought it beft to proceed. We did; and now began to perceive at a diftance fome little huts or hovels indiftinctly: a few more fteps affured us we were right, and about nine
o'clock

o'clock we had the good luck to find ourfelves at the very
hovel, where we were to reft that night. I own I now
found myfelf quite contented, though I did not at all
know what kind of place I was going to enter. It proved
to be a little hut made of boards, confifting of one apart-
ment only, eighteen or twenty feet fquare, and about
twelve high in the center, without any windows or
chimney for the fmoke, except what was made by the
holes in the roof, and the interftices between the boards
at the fides, which were rudely put together, fcarce
clofer than park-palings, affording an equal entrance to
the wind, rain, and fnow; for as thefe hovels are inhabited only for about four months in the fummer, they
are conftructed without the leaft mortar or cement in
the world; an humiliating witnefs this, how fimple the
architecture which nature and neceffity fuggeft. On en-
tering we found a comfortable fire, and the little *cabane*
inhabited by a couple of Alpine fhepherdeffes and their
two cows, on whofe whey and fome very coarfe bread
they wholly fubfifted, not difcontented but even proud of
their lot; and who, out of a fingular fpecies of contempt,
call the inhabitants of the plain *mange-rotis*, that is, eaters of
roaft-meat. Their language too was different; not French,
nor Italian, but partaking fomething of both; or, as I have
been fince informed, a corruption of the ancient Celtic.

5   A few

A few minutes after our arrival our guides rejoined us: it was now night, and in this rather too artleſs habitation we were obliged to lay in a little loft over the cows, our beds ſome leaves and clean hay, and my bolſter my port-manteau[n]. I had had the caution to bring ſome ſheets with me, and, being a little tired with my walking, ſlept five hours pretty ſoundly, though much ſtarved, having no other curtains than what this wooden canopy afforded, through which the ſtars ſhone moſt brilliantly. Between four and five we aroſe; found the heavens beautifully ſerene, and, having eaten ſome of our proviſions, left this habitation, which might be ſituated about two-thirds of the way up the mountain; and beginning our march about half after five reached the ſummit a quarter before ſeven; but not without a good deal of climbing, and ſometimes up an aſcent of near 40° for ſeveral hundred feet. One of my ſervants, before he got half way, found his head turn round, and himſelf ſo giddy, at the height and precipices (a frequent effect in theſe ſort of places) that he was obliged to return to the hut. In the aſcent I ſaw the Sun riſing behind one of the neigh-

(n) ——————————— Frigida parvas
Præberet ſpelunca domos, ignemque, laremque,
Et pecus, et dominos communi clauderet umbrâ;
Sylveſtrem montana torum cùm ſterneret uxor
Frondibus et culmo.   JUV. Sat. vi.

bouring

bouring alps with a most beautiful effect, and the shadow of the mountain we were then upon extended fifteen or twenty miles west. We had now reached the summit; and there my curiosity finished in astonishment. I perceived myself elevated 6000 feet in the atmosphere, and standing as it were on a knife-edge, for such is the figure of the ridge or top of this mountain; length without breadth, or the least appearance of a plain, as I had expected to find. Before me an immediate precipice, *à pic*, of above 1000 feet, and behind me the very steep ascent I had just now mounted. I was imprudently the first of the company: the surprize was perfect horror, and two steps further had sent me headlong from the rock.

On this spot, with some difficulty, we fixed the instruments, and commenced our operations, after some time spent in admiration at the prospect, and familiarizing myself to the scene. Before me, at some distance, was spread the plain in which lay Geneva and the lake; behind it rose the Dole, and the long chain of Mont Jura as far as the fort La Cluse, which we entirely commanded, as well as some of the country beyond it. A little to the left, and much nearer, lay Mont Saleve, which from this height appeared an inconsiderable hill: to the right and left nothing but immense mountains, and pointed rocks of every possible shape, and forming tremendous precipices. In the

vale beneath, feveral little hamlets, and the moft beautiful
pafturages, with the river Arve winding and foftening
the fcene; from whence arofe a thick evaporation, col-
lecting itfelf into clouds, which on the lake, that was
quite covered with them, had the appearance of a fea of
cotton, the Sun-beams playing in the upper furface of
them with thofe tints that are feen in a fine evening.
To the fouth-weft appeared the lake of Annecy; behind us, taking up one-fifth of our horizon, lay the
Glacieres, and amongft them, towering above all the reft,
ftood Mont Blanc. The circumference of the horizon
might be about 200 Englifh miles; and, though not one
of the moft extenfive, yet certainly one of the moft varied
in the world. From this fpot the clouds had a ftriking
appearance to an inhabitant of the plain; very few of
them at above one-fifth of the height that we were now
at; not governed by the wind, but moving in every poffible direction; fome of them feemed creeping along the
ground, whilft others were rifing perpendicularly between the hills. And I may here remark, that from
Geneva I have obferved the clouds were generally three
days in the week below the fummit of Mont Saleve; fo
that the ordinary region of thefe vapours feems to be at
that height in the atmofphere, where the barometer
would ftand at about 26 inches in this climate.

7

While

While at the top of the Mole, I was very sensible of the cold, there being a brisk wind, which, though south, came over the mountains of ice, and was very keen; insomuch that, about two hours after I had been there, I nearly lost the use of my fingers, and found my lips much affected and parched from the transition, having been a good deal heated in ascending with two waistcoats and a great coat on. The thermometer, however, when I first mounted, stood no lower than 48°. I must here ask pardon for this long digression, which I have ventured to transcribe from my journal written upon the spot.

To return then to the observations. After what has been said respecting those on Mont Saleve, it will suffice here to mention, that by repeated measurements I determined the horizontal length of the base 1, 2 (see the chart) to be = 1250 ft. 3.9 inch; the ∠ at 1 = 95° 37′ 28″; ∠ at 2 = 77° 48′ 53″; and the ∠ at 3 = 6° 33′ 49″. The mean corrected angle of elevation of 3 from 1 = 21° 29′ 34″; ditto of 3 from 2 = 21° 3′ 41″; and lastly, the elevation of 2 from 1 = 0° 47′ 24″.

These

Feet.

Thefe obfervations give for the length of the fide 1, 3,   —   10691.9

—   —   —   —   2, 3,   —   10886.7

Height of 3 above 1,   —   —   —   4212.8

——— 3 above 2,   —   —   —   —   4194.8

——— 2 above 1,   —   —   —   —   17.2

And confequently, 3 above 1 deduced from the obfervation at 2,   —   4212.0

And laftly, the mean height of 3 above 1 from the determination at } each end of the bafe,   —   —   —   } 4212.4

The difference in height, however, between the two barometers was only 4211.3 feet.

Here follow the barometrical obfervations [o], and their reduction.

(o) Made between the hours of eight and twelve, in the open air and not in the tent, which could not be pitched on accout of the fmallnefs of the plain at the fummit; a brifk fouth wind, but fair. The barometer was fcreened by an umbrella.

Comparifon of the firft feries on the Mole.

Obfervation at the top at 3.

	Barom. Nº 2. above at c. In. Pts.	Therm. attached.	Therm. detached.
	24.1437	57.0	54.8
Correct for the Diff. of the two attached therm. 3°.4, }	+ 88		
Barometer at the top,	24.1525	Log. 3829621	
———— below, —	28.1253	Log. 4490971	
Difference, or fall of the quickfilver, — }	3.9728	Diff. of Log. 661.350	{ approx. height in fathoms.
Correct for 18°.6 of heat, — —		+ 27.431	
Corrected height in fathoms, — —		688.781	
		× 6	
Height in feet by the barometer —		4132.686	
———————— by the geometrical meafurement,		4211.3	
Difference, or error of the barometer,		— 78.6 $= \frac{187}{10000}$.	

Obfervation below at 1.

	Barom. Nº 1. below at 1. In. Pts.	Therm. attached.	Therm. detached.
	28.1295	60.4	61.9
Correct for the diff. of barometer, }	— 42		54.8 heat at 3.
	28.1253		58.3 mean heat. 39.7 ftandard temperature.
			+ 18.6 difference.

Com-

Compariſon of the ſecond Series.

Obſervation at the top at 3.

	Barom. N° 2. above at 3. In. Pts.	Therm. attached.	Therm. detached.
	24.1420	56.9	56.0
Correct for the diff. of the two attached therm. 3°.5, }	+ 91		
	24.1511	Log. 3829369	
	28.1258	Log. 4491049	
Difference, or fall of the quickſilver, — }	3.9747	Diff. of Log. 661.680	{ approx. height in fathoms.
Correct for 19°.2 of heat, — —		+ 28.330	
Correct height in fathoms, — —		690.010	
		× 6	
Height in feet by the barometer, —		4140.06	
——— by the geom. method, —		4211.3	
Difference, or error of the barometer,		— 71.2 = $\frac{169}{1000}$.	

Obſervation below at 1.

	Barom. N° 1. below at 1. In. Pts.	Therm. attached.	Therm. detached.
	28.1300	60.4	61.8
Correct for the diff. of barometer, }	— 42		56.0 heat at 3.
	28.1258		58.9 mean heat.
			39.7 ſtandard temperature.
			— 19.2 difference.

Com-

Comparifon of the third Series.

Obfervation at the top at 3.

	Barom. N° 2. above at 3. In. Pts.	Therm. attached.	Therm. detached.
	24.1670	56.0	56.0
Correct for the diff. of the two attached therm. 4°9. }	+ 127		⊙ 57.5 (*p*)
	24.1797	Log. 3834509	
	28.1278	Log. 4491358	

Difference, or fall of the quickfilver, }	3.9481 Diff. of Log. 656.849	{ Approx. height in fathoms.
Correct for 19°8 of heat,	— — + 29.0	

Correct height in fathoms,   —   —   685.849
           × 6

Height in feet by the barometer,   —   4115.094
————— by the geom. method,   —   4211.5

Difference, or error of the barometer,   — 96.2 = $\frac{111}{10000}$.

Obfervation below at 1.

	Barom. N° 2. below at 2. In. Pts.	Therm. attached.	Therm. detached.
	28.1320	60.9	63.0
Correct for the diff. of the barometer, }	— 42		56.0 heat at 3.
	28.1278		59.5 mean heat.
			39.7 ftandard temperature.
			19.8 difference.

(*p*) In this column for the detached thermometer at the top of the mountain, in this and the following obfervations, are inferted two numbers; the upper one expreffing the heat in the fhade; and the lower one, with this mark ⊙ prefixed, the heat in the Sun. The computation, however, is made from the former; this may ferve to fhew the difference.

                                                  Com-

Compariſon of the fourth ſeries.

Obſervation at the top at 3.

	Barom. N° 2. above at 3. In. Pts.	Therm. attached.	Therm. detached.
	24.1780	57.2	56.0
Correct for the diff. of the two attached therm. 4°6, }	+ 119		O 57.5
	24.1899	Log.	3836341
	28.1318	Log.	4491976
Difference, or fall of the quickſilver, }	3.9419	Diff. of Log.	655.635 { approx. height in fathoms.
Correct for 20°.3 of heat,	—	—	+ 29 678
Correct height in fathoms, .	—	—	685.313 ×6
Height in feet by the barometer,	—		4111.878
———————— by the geom. method,	—		4211.3
Difference, or error of the barometer,			— 99.4 = $\frac{994}{10000}$.

Obſervation below at 1.

	Barom. N° 1. below at 1. In. Pts.	Therm. attached.	Therm. detached.
	28.1360	61.8	63.9
Correct for the diff. of the barometer, }	42		56.0 heat at 3.
			60.0 mean heat.
	28.1318		39.7 ſtandard temperature.
			+ 20.3 difference.

Com-

Comparifon of the fifth feries.

Obfervations at the top at 3.

	Barom. N° 2. above at 3. In. Pts.	Therm. attached.	Therm. detached.
	24.1840	59.6	57.0
Corre&t for the diff. of the 2 attached therm. 2°.8, }	+ 73		⊙ 59.3

	24.1913	Log.	3836592
	28,1308	Log.	4491820

Difference, or fall of the quickfilver, }	3.9395 Diff. of Log. 655.228 } approx. height in fathoms,
Corre&t for 20°.8 of heat, — —	+ 30.391
Corre&t height in fathom, — —	686.619 × 6
Height in feet by the barometer, —	4113.714
————— by the geom. method, —	4211.3
Difference, or error of the barometer,	— 97.6 = $\frac{811}{10000}$.

Obfervations below at 1.

	Barom. N° 1. below at 1. In. Pts.	Therm. attached.	Therm. detached.
	28.1350	62.4	64.0
Corre&t for the diff. of the barometer, }	— 42		57.0 heat at 3.
			60.3 mean heat.
	28.1308		39.7 ftandard temperature.
			+ 20.8 difference.

Compariſon of the ſixth ſeries.

Obſervation at the top at 3.

	Barom. N° 2. above at 3. In. Pts.	Therm. attached.	Therm. detached.
	24.1900	61.0	57.0
Correct for the diff. of the two attached therm. 1°6, }	+ 41		⊙ 60.0
	24.1941	Log. 3837095	
	28.1268	Log. 4491204	

Difference, or fall of the quickſilver, }	3.9327 Diff. of Log. 654 157	{ approx. height in fathoms.
Correction for 20°6 of heat,	— — + 30.048	
Correct height in fathoms,	— — 684.157 × 6	
Height in feet by the barometer,	— 4104.942	
————— by the geom. method	— 4211.3.	
Difference, or error of the barometer,	—106.4 = $\frac{858}{10000}$.	

Obſervation below at 1.

	Barom. N° 1. below at 1. In. Pts.	Therm. attached.	Therm. detached.
	28.1310	62.6	63.6
Correct for the diff. of the barometer, }	— 42		57.0 heat at 3.
	28.1268		60 3 mean heat.
			39 7 ſtandard temperature.
			20.6 difference.

To

To collect thefe laft experiments in one point of view.

				Feet.
The 1ft feries gives for the error on every 1000 ft.				18.7
2d,	—	—	—	16.9
3d,	—	—	—	22.8
4th,	—	—	—	23.5
5th,	—	—	—	23.1
6th,	—	—	—	25.2
The mean error,		—		21.7

which agrees within two feet in a thoufand with the determination on Mont Saleve. This refult then juftifies my conclufion (in p. 556.) and proves that either the proportional gravity of air and quickfilver is now different from what it was, when M. DE LUC made his experiments, *viz.* from 1756 to 1760; or that his or my obfervations are defective. That my trigonometrical meafurements were fufficiently exact, *viz.* to within two or three feet, I think I have already fhewn; and even that his were alfo. Within what limits my barometrical errors are to be found is not difficult to determine from what has been before premifed. That the fcale of Mr. DE LUC's barometer was lefs accurate than mine, is, I think, without a doubt; and indeed he never attempted a divifion lefs than $\frac{1}{16}$th of a French line, or about $\frac{1}{1000}$

of

of an inch Engliſh: and yet when I conſider the number of his obſervations, and the unexampled diligence and care with which he made them, I am obliged to attribute the difference of our reſults to ſome other cauſe than that of inaccuracy. If then future experience ſhould demon-ſtrate, that the denſity of the atmoſphere with a given heat is invariable, or nearly ſo; while the preſſure of a whole column of it continues the ſame, we may perhaps ſearch for the cauſe of our diſagreement from hence, *viz.* the barometers of Mr. DE LUC were not ſufficiently near each other in an horizontal direction: mine were ſeparated from two to three miles; and his, I believe, at double or triple that diſtance. It may be ſuſpected, I am well aware, that the ſyphon conſtruction of Mr. DE LUC's barometer might occaſion this difference: let us ſee whe-ther this be the caſe. Mr. DE SAUSSURE (whoſe inſtru-ment was of Mr. DE LUC's conſtruction, and made, as I underſtood, under his inſpection) obſerved at the top of the Mole, or at leaſt nearly on the ſame level with my barometer, as follows:

And

	Barometer In. L. 16ths.	Therm. attached. DE LUC's scale.	Therm. det. REAUM scale.
	22 8 0	+1°+	+10°—
And in English measure and FAHREN-HEIT's scale, — —	24.1570	56	54.2
Mr. DE SAUSSURE's barometer ordinarily stands higher than mine. N° 2. by (1),	—.0117		
Correct for the diff. of our attached therm. 1°,	+ 26		
Mr. DE SAUSSURE's barometer corr ted,	24.1479		
My barometer N° 2. see the first series,	24.1437—	57	54.8
Difference, — —	+.0042 wholly inconfiderable.		

Our barometers may therefore be said to have agreed exactly.

Mr. DE SAUSSURE made a second comparison just before we left the top of the mountain, which proved as follows.

	Barometer In. L. 16ths.	Therm. attached. DE LUC's scale.	Therm. detached.
	22 8 8	+4°	+11$\frac{20}{3}$
Or reduced to English measure and scale,	24.2014	61.7	57.9
Mr. DE SAUSSURE's barometer stands higher than mine N° 2. —	—.0117		
Corr. for the diff. of our attached therm. 0°.7,	—.0018		
Mr. DE SAUSSURE's barometer corrected,	24.1879		
My barometer N° 2. see the sixth series,	24.1900	61.0	57
Difference, — —	—0.0021		

So that, in the first comparison, his barometer at the top of the Mole stood higher than mine by +,004 inch; and in the last, lower by —,002; the mean is higher by

(1) This we found by comparisons at the bottom of the mountain.

+,001

I

+,001, equal to about 10 inches in deducing the height of the mountain, a quantity wholly to be neglected. Finally, the mean of Mr. DE SAUSSURE's obſervations gives the defect of Mr. DE LUC's rules 21.9 in a thouſand. The conſtruction of the barometer had therefore no influence on this difference. But further, while Mr. DE SAUSSURE obſerved the height of the barometer on the Mole, Mr. DE LUC, the brother made a correſponding obſervation with a ſimilar inſtrument at Geneva. I ſhall relate this obſervation, computed after Mr. DE LUC's manner.

Mr.

In. L. 16ths.

Mr. DE SAUSSURE, at 4 feet below the summit of the Mole, —	22	8	0

Mr. SAUSSURE's barom. stands higher than Mr. DE LUC's ordinarily by,	+	1½

Heat of the air.

Thermometer attached + 1°,   — 0¼

16ths of a line.   Log.   REAUM. DE LUC'S   Therm. Therm.

Correct height on the Mole,   22   8   0¾ = 4352.¾   6387587   +10   —15¾

Mr. DE LUC, 78 feet above the lake, —   } 27   0   0

Therm. attached +6°,   —   — 6

26   11   10 = 5178   7141620   +15   — 4

Difference of the Log.   —   —   754.033   Sum —19¾

$19\frac{°1}{3} \times \frac{110411}{1000}$ = the correction for the temperature,   —14.854

Correct height in French toises,   —   739.179
   × 6

Height in French feet,   —   —   4435.074
Mr. DE LUC's barometer above the lake of Geneva,   +78.
Mr. SAUSSURE's barometer below the summit of the Mole,   —   —   —   } + 4

And consequently, the summit of the Mole above the lake, in French feet,   —   —   } 4517.

Which reduced to English feet is,   —   —   4814.

But, by a mean of my trigonometrical operations, this height is (*vide* chart)   —   —   } 4883.

Difference, or error of the barometrical rules, $-69. = \frac{1}{1000}$

This last observation serves at least to shew, that the error I am contending for is on the defective side, though it gives the quantity of it somewhat less, but by no means deserves that confidence which the other comparisons do; for, besides that this single observation may be concluded

less

lefs decifive, the trigonometrical meafurement is alfo lefs accurate from the diftance; and, laftly, to fuppofe the ftate of the atmofphere precifely the fame with refpect to weight in two places twenty miles afunder, is, I am afraid, a *poftulatum* too hazardous to grant. I therefore fay, that all thefe obfervations confirm the fame truth, that the atmofphere is lighter than Mr. DE LUC prefumed it. What had already been done may feem fufficient for the eftablifhment of this fact; for I have always held, that a few obfervations, well made and faithfully related, do more in the interpretation of nature, than a multitude of crude, carelefs, and immethodical experiments. But I have not done: I wifhed to put this matter out of all doubt, and accordingly undertook another expedition to the fummit of Mont Saleve, on the 18th of September, and in a colder temperature: the experiments then made, with their refults, were as follows :

The difference of actual height by the two barometers was 2828.9 feet, the barometer N° 1. ftanding higher than N° 2. by +,0038 inch, when compared at the bottom of the mountain.

Com-

Comparison of the first series.

Observation at the top of the mountain.			Observation at the bottom.		
Barom. N° 2. at the top. In.	Therm. attached.	Therm. detached.	Barom. N° 1. below. In.	Therm. attached.	Therm. detached.
25.6533	58.0	56.2	28.4040	58.1	58.8

	Feet.
This gives for the height barometrically,	2755.6
But the true height was, —	2828.9

Difference, or error of the barometers, $-73.3 = \frac{253}{10000}$.

Comparison of the second series.

Observation at the top of the mountain.			Observation at the bottom.		
Barom. N° 2. at the top. In.	Therm. attached.	Therm. detached.	Barom. N° 1. below. In.	Therm. attached.	Therm. detached.
25.6550	56.2	57.0	28.4040	58.5	60.8

	Feet.
This gives for the height barometrically,	2754.9
But the true height was, —	2828.9

Difference, or error of the barometers, $-74.0 = \frac{262}{10000}$.

Compariſon of the third ſeries.

Obſervation at the top of the mountain.			Obſervation at the bottom.		
Barom. N° 2. at the top. In.	Therm. attached.	Therm. detached.	Barom. N° 1. below. In.	Therm. attached.	Therm. detached.
25.6620	56.2	57.2	28.4040	59.3	62.0

Feet.

This gives for the height barometrically, 2748.9
The height by the trigon. method was, 2828.9

Difference, or error of the barometers, $-80.0 = -\frac{282}{10000}$.

Compariſon of the fourth ſeries.

Obſervation at the top of the mountain.			Obſervation below.		
Barom. N° 2. at the top. In.	Therm. attached.	Therm. detached.	Barom. N° . below. In.	Therm. attached.	Therm. detached.
25.6600	56.4	57.4	28.4040	59.3	62.2

Feet.

This gives for the height barometrically, 2752.8
But the true height was, — 2828.9

Difference, or error of the barometers, $-76.1 = \frac{169}{10000}$.

In theſe compariſons I have not inſerted the whole of the computation, as that may eaſily be made by any perſon at leiſure. Finally, the mean of theſe four laſt ſeries feries gives for the error on 1000 feet, 26.8. I think I have now fhewn, that the error actually exifts; it remains that we determine precifely the quantity of it. For this purpofe it will be proper to collect all the preceding obfervations in one point of view.

Table of the refult of all the barometrical experiments.

Place of obfervation.		True height trigonometrically.	Height by the barometers.	Mean heat.	Error in feet.	Error in 1000 feet.	Mean error in 1000 feet.
Mont Saleve,	1	2831.3	2775.2	69.4	— 56.1	—19.8	
	2	———	2763.2	68.5	— 68.1	—24.0	} —23.1
	3	———	2759.4	67.2	— 71.9	—25.4	
At the Mole,	1	4211.3	4132.7	58.3	— 78.6	—18.6	
	2	———	4140.1	58.9	— 71.2	—16.9	
	3	———	4115.1	59.5	— 96.2	—22.8	
	4	———	4111.9	60.0	— 99.4	—23.5	} —21.7
	5	———	4113.7	60.5	— 97.6	—23.1	
	6	———	4104.9	60.3	106.1	—25.2	
Mont Saleve,	1	2828.9	2755.6	57.5	— 73.3	—25.9	
	2	———	2754.9	58.9	— 74.0	—26.2	} —26.8
	3	———	2748.9	59.6	— 80.0	—28.2	
	4	———	2752.8	59.8	— 76.1	—26.9	

Mean of all, 23.6, and the temperature 61°.4.

The Mole, from two obfervations of Mr. DE SAUSSURE, —		4211.3	———	—	— 92.	—21.8	
The fame by Mr. DE SAUSSURE, and Mr. DE LUC, at Geneva,		4883.	4814.	—	— 69.	—14.	} —16.2
According to Mr. DE LUC's own obfervations, fee *Recberches fur l'atmofphere,*	the Mole,	4882.8	4860.	—	— 22.8	— 47	
	the Dole,	4292.7	4210.	—	— 82.7	—19.5	
	the Buet,	8893.6	8770.	—	—123.7	—13 9	
	Mt Blanc,	14432.5	14093.	—	—339.5	—23.5	

The

The titles of the columns are sufficiently clear to make a farther explanation of this table unneceffary; and it appears, I think inconteftably, upon taking a mean of my thirteen obfervations (and I fhall here confider only my own) on Mont Saleve and the Mole, that this error is about $23\frac{1}{4}$ feet on every thoufand; that is, the rules of Mr. DE LUC give the height by fo much too little. At the bottom of the foregoing table I have fubjoined fix other comparifons, fome of them from Mr. DE LUC's own obfervations, as recorded in his valuable work; which however I muft add, are certainly of lefs authority in this inquiry, as they were made with barometers a great way diftant from each other, *viz.* near thirty miles: befides which, the geometrical heights are, for the fame reafon, not fo accurately afcertained. I have, however, ventured to make what ufe I could of them, *viz.* to fhew that thefe two give a refult on the fame fide, though not exactly the fame; and to urge the neceffity of a certain vicinity in thofe obfervations from whence a theory is to be deduced.

Shall I be permitted to adduce another proof, in confirmation of what has been advanced? When I firft took up the confideration of meafuring altitudes in the atmofphere with the barometer, and had heard only of Mr. DE LUC's labours, it occurred to me, that there was a

6   much

much more fimple method of arriving at this theory, than either he or I have fince purfued. It was this; to determine hydroftatically the fpecific gravities of air [r] and quickfilver, with a given temperature and preffure; the increafe of volume, or change of gravity, with a given increafe of heat being fuppofed to be known by the experiments of BOERHAAVE [s] and HAWKESBEE [t], which might be farther examined by fimilar ones; and prefuming that the geometrical ratio in the air's denfity, as you advance upwards from the earth's furface, had been fufficiently demonftrated [u]. For the proportional gravity of quickfilver to air will exprefs inverfely the length of two equiponderant columns of thefe fluids, that is, when the columns are taken infinitely fmall [x]. With thefe

(r) It may feem particular that I fhould propofe an experiment fuppofed to be very well known, and which hardly any elementary treatife on chemiftry or experimental philofophy will not furnifh us with an example of; the weight of a given quantity of air. BOYLE, HALLEY, HAWKESBEE, HALES, each of them have tried it, and many others fince their time: but the misfortune is, all thefe experiments have been but grofs approximations, without due attent on to the heat; and yet the determination of HAWKESBEE feems to have been followed by one-half of Europe in Pneumatical refearches. Indeed I only know of one experiment that has the leaft title to precifion, and that is Mr. CAVENDISH's, briefly related in the LVIth volume of the Philofophical Tranfactions.

(s) Elementa Chemiæ.

(t) Phyfico-mechanical Experiments.

(u) COTES's Hydroftat. Lectures, *et alibi.*

(x) I am not forry to anticipate the reader's remark here, that this obfervation is not new; fince I find that I have been treading the fame fteps with

Mr.

thefe ideas I made the following experiment. I caufed a glafs veffel to be blown fomething like a Florence flafk, or rather larger; to the neck of this was adapted a brafs cap with a valve opening outwards, and made to fcrew on or off, together with a male fcrew, by which it was fixed to an excellent pump of Mr. NAIRNE's conftruction, and exhaufted of its air, or at leaft rarified to a known degree: the veffel was then carefully weighed with a fenfible balance, and again after the air was re-admitted; the difference gave the weight of the air that had been exhaufted. After having repeated this two or three times, the veffel was exactly filled with water as far as the valve, which had been the term of capacity for the air; this was done by fcrewing on the cap till the fuper-fluous water oozed all out, and upon inverting the veffel there appeared not the leaft fign or bubble of air; I therefore concluded, that the volume of water was pre-cifely the fame as had been the volume of air, a circum-ftance that fhould be accurately attended to. It was then carefully weighed, and compared with its weight when full and deprived of its air. It will readily be feen, that I had then the fpecific gravity of the two fluids, upon fuppofition that the figure of the glafs had not altered

Mr. BOYLE and Dr. HALLEY, who both made ufe of this method; the one with a view to determine the limits of the atmofphere; and the other the height of Snowden.

2   **by**

by pressure during the experiment; and this effect may be presumed to have been the most sensible, when the vessel was filled with water, the pressure at that time being from within. To assure myself of this, I let in a small quantity of air, which formed a bubble of about one-third of an inch in diameter, and upon immerging the glass in another vessel of water, whereby the pressure within was counterpoised by a pressure without, the bubble seemed to contract itself by a quantity, as I found afterwards, equal to about two grains in weight, or $\frac{1}{8000}$ of the whole contents. I therefore concluded, that this correction was hardly worth taking notice of, and still less the effect from external pressure when the glass was exhausted. At every operation the height of the barometer and thermometer (placed close to the vessel when the air was weighed) was noted down, together with the height of the pump-gage, which, compared with the barometer in the room, shewed the quantity exhausted. The result of the experiment was as follows, the barometer in the room standing at 29.27 inches, and the heat of the room 53°.

The

	Feet.
The bottle empty or exhausted till the gage stood at 29.15 inches weighed (determined from four different trials, and the balance turning with $\frac{1}{12}$ of a grain) — — —	2657.40
Increase of weight when filled with air, from four trials certain to $\frac{1}{16}$ of a grain — — —	+16.13
Bottle filled with water, whose heat was 51°, — —	16220.00
Weight of the water, exclusive of the bottle, —— —	13562.60

But the bottle was exhausted only in the proportion
of 29.15 inches to 29,27 inches; therefore if a perfect
vacuum could have been made, the difference of weight
would have been 16.22 grains instead of 16.13 grains.
Again, the water was colder than the air by 2°; the one
being 53°, and the other only 51°: now water, from
former experiments, I find to expand about $\frac{1}{10000}$ with
2° of heat; therefore, if the water had been of the same
temperature with the air that was examined, the weight of
an equal volume would have been only 13558,5 grains;
and lastly, 13358.5 divided by 16.22 gives 836 [y], and
by so much is water heavier than air in these circumstances.

[y] HAWKESBEE's experiments made the air 850 lighter than water, the barometer being at 29.7; and Dr. HALLEY supposed it about 800. Mr. CAVEN-
DISH, in weighing 50 grains of air, when the barometer was at 29$\frac{1}{2}$, and the
thermometer at 50°, concluded the specific gravity of air to be about 800 also.
Now my experiment, reduced to the same circumstances with his, would give
817 for this gravity, no great difference in an affair of such delicacy.

By

By former experiments I find the specific gravity of the quicksilver of my barometers, compared with rain-water in 68° } 13.606 to 1
of heat, as,   —   —   —   —

And 68°—53°=15°, correct therefore for 15° of expansion of } + .018
quicksilver,   —   —   —

Correct for 15° of expansion of air,   —   —   —.031

True specific gravity of quicksilver, with 53° of heat,   13.594

Which multiplied by the specific gravity of air,   —   × .836

Gives for the comparative gravity of quicksilver and air, when } 11364.6
the barometer is at 29.27, and the thermometer 53°,   —

                                                             Feet.

And lastly, $\frac{1}{100}$th of an inch of quicksilver, when the barometer stands at }
29.27 inches (*viz.* from 29.22 inches to 29.32 inches) with the tem- } 94.7
perature 53°, is equal to a column of the atmosphere of,   —

This quantity, according to my barometrical observations, is,   —   93.83

—————————————— to Mr. DE LUC's rules,   —   91.66

We see here then that the statical experiment agrees
with the result of my barometrical ones to within about
11 inches in 100 feet, and I am not sure that it is not
still capable of much farther precision; and though perhaps alone it might carry with it, to some persons, a less
conclusive testimony, who reluctantly reason from the
little to the great, yet, in conjunction with what has been
before shewn, I think it has considerable weight; and I
am the less inclined to reject such an indirect method of
proof, as I have the great authorities of HALLEY and
NEWTON on my side[x].

                                                        I have

(x) " Ce qu'il y a d'essentiel à observer ici," says Mr. DE LUC, " et vrai-
" ment digne de remarque, c'est que par la seule connoissance des pesanteurs
" specifiques de l'air et du mercure, HALLEY est parvenu à une regle très

VOL. LXVII.   4 D   " approchante

I have thus endeavoured to fhew then that the error of the theory is $- \frac{216}{10000}$ when the temperature of the air is 61°.4 (fee the table of the refult of the obferva-tions). It remains therefore, finally, that we deduce a rule, the error of which fhall be nothing, *viz.* to find the temperature wherein the difference of the loga-rithms of the heights of the barometer, taken to four places of figures, will give the true difference of eleva-tion in Englifh fathoms. Previous to this invefligation, with which I intend to conclude this paper, it will be ne-ceffary to remark, that by repeated experiments with the barometer, I find a fmall difference in the equation for the expanfion of air by a change of temperature, and even in that of quickfilver from the fame caufe, from what Mr. DE LUC's obfervations have given it[a]. I fhall

"approchante de celle, qu'un grand nombre d'obfervations du baromètre dans
"les Cordelières ont dicté depuis à M. BOUGUER: cependant malgré l'appui
"que ces experiences fe prêtent reciproquement, on verra qu'elles atoient encore
"bien eloignées de fournir une regle generale." Recherches fur l'Atmofphere,
fect. 267.

(a) He indeed made his experiments on the atmofphere itfelf with the barometer, in order to determine the variations of its denfity; but fince it appears that the abfolute denfity of this fluid is different from what he fup-pofed it, it is no bold conjecture to prefume that the degree of its variation fhould be different alfo; and to afcertain this point, I have preferred the inftrument above-mentioned to the method ufed by Mr. DE LUC, how direct foever his may feem; for in determining minute quantities or equations, we muft not embarrafs ourfelves with the compound effect of too many caufes at a time.

not

not here trouble the reader with the experiments at large, too fimple in themfelves to deferve fuch a detail, unlefs a future occafion fhould render that neceffary, as the methods here ufed may be met with amongft HAWKESBEE's or Mr. BOYLE's experiments; and content myfelf with relating only the refult of the different trials.

1000 parts of air of the temperature of freezing and preffure of $30\frac{1}{2}$ inches, increafed in volume by an addition of 1 degree of heat on FAHRENHEIT's thermometer as follows:

Obfervations.		Number of degrees the air was heated.	Expanfion for 1° in 1000ths of the whole.	
With the firft manometer,	1	14.6°	2.30	Mean from the firft manometer 2.44.
	2	32.2	2.43	
	3	40.3	2.48	
	4	46.6	2.45	
	5	49.7	2.48	
	6	51.1	2.51	
	7	23.7	2.36	
	8	13.1	2.24	
With another manometer,	9	22.0	2.38	Mean from the fecond manometer 2.42
	10	28.0	2.50	
	11	21.5	2.34	
	12	30.1	2.44	
	13	22.6	2.44	

The

The mean of theſe two ſorts of obſervations, made with different inſtruments, is 2.43, *viz.* 1000 parts of the air at freezing become by expanſion from 1° of heat

Pts.   Pts.

equal 1002.43 or 1002.385 with the ſtandard temperature 39°.7. Mr. DE LUC's experiments reduced give

Pts.

this quantity equal 1002.23[b] (ſee Tranſ.). It may be imagined, that I ſhould have had a more accurate concluſion by making theſe obſervations in greater differences of temperature than what is ſhewn in the ſecond column of the above table; but it did not appear ſo to me. On the other hand, I found that it was abſolutely neceſſary that the ſame heat ſhould be kept up for ſome hours together, in order that I might be ſure that the air within the inſtrument, the glaſs tube that contained it, and the air without it, all had acquired the ſame

---

(b) It has generally been ſuppoſed, that air expands $\frac{1}{100}$ with each degree of the thermometer, commencing from the mean temperature 55°; and, in conſequence of this, aſtronomers have computed tables for correcting their mean refractions; but, upon reducing the reſult of my obſervations to the temperature 55°, we ſhall have the correction of the refraction for 1° = $\frac{10}{10000}$ or $\frac{1}{13}$. Now according to Mr. DE LUC this equation is $\frac{10}{1000} = \frac{1}{167}$, which would produce a difference of about 4″ in the corrected refraction, upon an altitude of 5°, with the temperature 35°. If my numbers may be ſuppoſed to deſerve equal confidence, the error of the tables in common uſe, in the above circumſtances, would amount to only half that quantity, and therefore probably will be thought ſcarce worth correcting. I have mentioned this in order to obviate the concluſions that have been drawn by ſome perſons from Mr. DE LUC's theory.

uniform

uniform temperature, which in my room I found not very eafy to effect in heats greater than 70° or 80°. I have therefore preferred repeating the experiment with fmall differences of heat; but fuch, however, as will include almoft all the temperatures in which barometrical obfervations are likely to be made, *viz.* from 32° to 83°.

It has been fufpected, in confequence of fome experiments made by a very ingenious member of this Society, that air does not expand uniformly with quickfilver; or that the degrees of heat fhewn by a quickfilver-thermometer would be expreffed on a manometer, or air-thermometer, by unequal fpaces in a certain geometrical ratio. I do not deny this propofition; but I have alfo very little reafon to affent to it, if I may truft my own experiments, which certainly evince that this ratio, if not truly arithmetical, is fo nearly fo as to occafion no fenfible error in the meafuring of heights with the barometer; and that is all I contend for. The fmall differences that are feen in the above table of this expanfion, deduced from a mean of 14° or of 40°, I would attribute rather to the errors of obfervation than to any actual irregularity in nature. If, however, this progreffion be infifted upon, it fhould feem, that the degree of the air's expanfion increafes with an increafe of heat; and that the difference of volume or denfity from 1° of heat,

heat, any where within the limits above-mentioned, would be about one part in five thouſand from what I take it at a mean. I ſhould not have inſiſted ſo long on this circumſtance, but in reſpect to the known accuracy of the author of this hypotheſis. Neither do I find any reaſon to believe, that the expanſion of air varies with its denſity. I have tried air whoſe denſity or preſſure was equal to 23¾ inches, and alſo to forty inches; but the dilatation, with equal volumes and equal degrees of heat, was very nearly the ſame in both caſes. I might add a great deal more on theſe manometrical experiments, but I am afraid it would be more tedious than uſeful. I proceed therefore to the expanſion of quickſilver.

This experiment was made with a tube, ſomething like a thermometer, but conſiderably larger than the ordinary ſize, and open at one end; it was filled with quickſilver to a certain height, and then expoſed to the temperatures of freezing and boiling repeatedly, the barometer being at 30 inches: the difference of the volume in each inſtance was determined afterwards by accurately weighing the contents. I thus found, that if the quickſilver at freezing be ſuppoſed to be divided into 13119 parts, the increaſe of volume by a heat of boiling water became equal to 208 of theſe parts = $\frac{10}{637}$, and $\frac{10}{637} \times \frac{1}{180} = \frac{1}{11466}$; and ſuch would be the expanſion for

2   each

each degree of the thermometer, commencing from the freezing point, $= 0,00262$ inch on a column of 30 inches of the barometer, if the glass had suffered no expansion during the experiment. This, however, has been found to be with 180° of heat $= \frac{1}{400}$ in solidity *(viz.* the cube of its longitudinal expansion) and $\frac{1}{400} \times \frac{1}{180} = \frac{1}{71000} = 0,00042$ inch, for the effect of the expansion of the glass for 1° upon a column of 30 inches; this added to the quantity before found, which was only the excess of the greater expansion above the less, gives for the true equation for each degree 0,00304 inch when the barometer stands at 30 inches[c]. Mr. DE LUC's correction in this case was 0,00312; a difference so small that I shall take no notice of it as to its influence upon our observations. It may deserve a remark here, that this equation rigorously taken is variable according to the height of the thermometer; for 1°, which at

[c] It has been suspected, and I believe will appear from very good observations, which however I never made myself, that the expansion of quicksilver in the barometer is not directly as the heat shewn by the thermometer, but in a ratio something different, owing to some of the quicksilver being converted into an elastic vapour in the *vacuum* that takes place at the top of the Torricellian tube, which presses upon the column of quicksilver, and thus counteracts in a small degree the expansion from heat. It does not, however, appear to be a considerable quantity, not amounting to above one-sixteenth of the whole expansion in a range of 40° of temperature; I shall therefore venture to consider this equation as truly uniform, since the error on ten thousand feet would not amount to five.

freezing

freezing is $= \frac{1}{9897}$ of the whole volume, at the temperature 82° becomes $\frac{1}{9941}$, a difference indeed that may
fairly be neglected, and which I neglect myſelf; yet I
cannot help obſerving, in juſtice to Mr. DE LUC, that his
method of reducing his barometers always to the ſame
ſtandard temperature, was free from the error I am
ſpeaking of.

To conclude, the defect of Mr. DE LUC's rules being
ſuppoſed $\frac{236}{100000}$, or, which comes to the ſame thing, the
correction being $+ \frac{2417}{100000}$, when the temperature of the
air is 61°.4, and the true expanſion of the air for each
degree being $\frac{239}{100000}$ when the heat is 39°.7; required to
find the temperature wherein the difference of the logarithms ſhall give the true height in Engliſh fathoms,
that temperature, according to Mr. DE LUC, being 39°.74,
and the expanſion $\frac{223}{100000}$.

Let T be the temperature 61°.4; s Mr. DE LUC's
ſtandard temperature; E the expanſion for 1°; e the ſame,
according to Mr. DE LUC; α the ſuppoſed correction of
the rules, and x the temperature ſought. We have then
the following formula, $\overline{T-s} \times \overline{E-e}^{(d)} - \alpha = s - x$, wherein
proceeding with the above numbers $s - x$ comes out

(d) This ſign is negative, becauſe the aſſumed expanſion *e* is leſs than the
true one E, and conſequently tended to increaſe the apparent error of the rules;
had it been greater, *e* would have been +.

$$8°.50$$

8°.50, and consequently $x=31°.24$ the temperature required; which, if it should be thought convenient, may be considered as the freezing point.

In the whole of the above inquiry I have taken no notice of the effect of gravity upon the particles of the air at different distances from the earth's center, which should doubtless enter into the account, and which would occasion the density of the atmosphere to decrease in a ratio something greater than the present theory admits of. In a height of four English miles Dr. HORSLEY finds (Phil. Transf. vol. LXIV.) that the diminution of density or volume from the accelerative force of gravity would be only $\frac{1}{500}$ part of the whole, or about 48 feet; and I may add to this, that this effect will be in the duplicate ratio of the heights, so that at one mile high it becomes only three feet. A like effect takes place also below the surface of the earth, as in measuring the depths of mines, &c. with this difference, that here it is but half the quantity; in the former instance gravity within the earth being simply as the distance from the center; they are both of them, however, circumstances that deserve no attention in practice.

This would be the place for me to enumerate the many desiderata, besides those already hinted at, that still remain for the perfection of this theory; such as the

laws of heat, that obtain in the different regions of the
atmoſphere; the effects of moiſture, winds, the electric
fluid, together with the weight and qualities of the air in
different countries, &c.; that at the ſame time that I am
congratulating the preſent age on one of the moſt bril-
liant diſcoveries in natural philoſophy, I may be underſtood alſo to encourage every lover of ſcience to ſtill farther
enquiries in a branch of knowledge no leſs uſeful than
ingenious; particularly in a kingdom wherein, from its
commercial intereſts, and in conſequence its many inland
navigations, every improvement in hydroſtatics, the art
of levelling, or geometry, cannot but tend conſiderably
to the public benefit.   The ſources of ſcience are not
eaſily exhauſted; multitudes of them remain wholly
unexplored.   If novelty can afford a charm, the path I
am ſpeaking of, till of late, has been the leaſt frequented;
witneſs the freſh, important truths in Pneumatical re-
ſearches that, from zeal and faſhion, every day's experience affords.   I might here offer too a tribute of applauſe
(and am ſure in concert with this whole aſſembly) juſtly
due to the indefatigable labours of him whoſe ſteps I
have purſued; but I am convinced he will rather hear
me acknowledge our obligations to the ancients than any
panegyric of himſelf.   Be the benefit we receive from
them our encouragement to proceed.

7   *Multum*

Multum egerunt, qui ante nos fuerunt, fed non pere-
gerunt: multum adhuc reflat operis, multumque reflabit;
nec ulli nato poft mille fæcula præcludetur occafio aliquid
adhuc adjiciendi." SEN. Epift. 64.

———————————

P A R T   II.

IN the fubfequent pages, which I have now the ho-
nour of laying before the Royal Society, I have drawn
up, and I hope in a form the moft commodious, the neceffary tables and precepts for calculating any acceffible
heights or depths from barometrical obfervations, and
without which I thought the preceding memoir would
be incomplete; referring, however, to that for the proofs
or elements from whence the tables have been computed. And herein I have endeavoured to adapt myfelf
to the capacity of fuch perfons as are but little conver-
fant with mathematical computations, but who may have
frequent opportunities of contributing fomething to the
advancement of fcience by experiments with this ufeful

inftruinſtrument, which is now become nearly in as common poſſeſſion as a pocket watch. I have induſtriouſly avoided the method of logarithms, propoſed by Dr. HAL-LEY, and adopted by Mr. DE LUC, both becauſe ſuch tables are not in the hands of every body, and becauſe I have perceived that many perſons of a philoſophical turn, though ſkilled only in common arithmetic, have been frightened by the very name: a method leſs popular, however elegant, would have been leſs generally uſeful. To theſe tables is ſubjoined a liſt of ſeveral altitudes, as determined by the barometer: this will ſerve to ſhew the uſe I have made of the inſtrument, and will at the ſame time exhibit the level of a great number of places in France, Savoy, and Italy, and, as I think, be no improper ſupplement to exemplify the rules. It might have been expected that I ſhould have ſaid ſomething on the theory of barometrical obſervations, and have laid down the laws and principles on which it depends; but as that has been ſo amply done by other writers of inconteſted authority, I ſhall content myſelf with inſerting only the following propoſitions.

1ſt, The difference of elevation of two places may be determined by the weight of the vertical column of the atmoſphere intercepted between them.

2   2d, If

2d, If then the weight of the whole atmofphere at each place can be afcertained, the weight of this column, *viz.* their difference, will be known.

3d, But the height of the quickfilver in the barometer expreffes the total weight of the atmofphere in the place of obfervation; the difference, therefore, of the height of the barometer, obferved in two places at the fame time, will exprefs the difference of elevation of the two places.

4th, But further, the weight of this column of the atmofphere is liable to fome variations, being diminifhed by heat, and augmented by cold; and again, a fimilar alteration takes place in the column of quickfilver, which is the meafure of this weight.

5th, If then the degree of thefe variations can be determined, and the temperature of the air and quickfilver at the time of obfervation be known, the weight of this column of air, or the difference of elevation of the two places, will be concluded as certainly as if the gravity of thefe two fluids, with all heats, remained invariably the fame: this is the whole myftery of barometrical meafurement.

APPLICATION.

The height of the barometer in Englifh inches at any two places at the fame inftant, and the heat (according to FAHRENHEIT's thermometer) to which it is expofed, being known, together with the temperature of the air at each place, obferved with a fimilar inftrument; required the difference of elevation of the two places in Englifh feet.

RULE.

Precept the 1ft, With the difference of the two thermometers that give the heat of the barometer (and which, for diftinction fake, I fhall call the attached thermometers^(√)) enter table I. with the degrees of heat in the column on the left hand, and with the height of the barometer in inches, in the horizontal line at the top; in the common point of meeting of the two lines will be found the correction for the expanfion of the quickfilver

(√) It is fcarce neceffary to remark, that, in order to make good concluſive obfervations, it is proper to be furniſhed with two barometers, and four thermometers; *viz.* one attached or inferted in the frame of each barometer; and the other two detached from them, in order to take the heat of the open air; for it will feldom be found, that the thermometer in the frame of the barometer and that in the air will ſtand at the fame point, and for a very evident reafon.

by

by heat, expreffed in thoufandth parts of an Englifh inch; which added to the coldeft barometer, or fubtracted from the hotteft, will give the height of the two barometers, fuch as would have obtained had both inftruments been expofed to the fame temperature.

Precept the 2d, With thefe corrected heights of the barometers enter table II. and take out refpectively the numbers correfponding to the neareft tenth of an inch; and if the barometers, corrected as in the firft precept, are found to ftand at an even tenth, without any further fraction, the difference of thefe two tabular numbers (found by fubtracting the lefs from the greater) will give the approximate height in Englifh feet. But if, as will commonly happen, the correct height of the barometers fhould not be at an even tenth, write out the difference for one entire tenth, found in the column adjoining, intitled *Differences*; and with this number enter table III. of proportional parts in the firft vertical column to the left hand, or in the 11th column, and with the next decimal following the tenths of an inch in the height of the barometer *(viz.* the hundredths) enter the horizontal line at the top, the point of meeting will give a certain number of feet, which write down by itfelf; do the fame by the next decimal figure in the height of the barometer *(viz.* the thoufandths of an inch) with this difference,

ftriking

ſtriking off the laſt cypher to the right hand for a fraction; add together the two numbers thus found in the table of proportionable parts, and their ſum ſubduct from the tabular numbers juſt found in table II.; the differences of the tabular numbers, ſo diminiſhed, will give the approximate height in Engliſh feet.

Precept the 3d, Add together the degrees of the two detached or air-thermometers, and divide their ſum by 2, the quotient will be an intermediate heat, and muſt be taken for the mean temperature of the vertical column of air intercepted between the two places of obſervation: if this temperature ſhould be $31°\frac{1}{4}$ on the thermometer, then will the approximate height, before found, be the true height; but if not, take its difference from $31°\frac{1}{4}$, and with this difference ſeek the correction in table IV. for the expanſion of air, with the number of degrees in the vertical column on the left hand, and the approximate height to the neareſt thouſand feet in the horizontal line at the top; for the hundred feet ſtrike off one cypher to the right hand; for the tens ſtrike off two; for the units three: the ſum of theſe ſeveral numbers added to the approximate height, if the temperature be greater than $31°\frac{1}{4}$, ſubtracted if leſs, will give the correct height in Engliſh feet. An example or two will make this quite plain.

E X A M P L E

EXAMPLE I.

Let the height of the barometer, obferved at the bottom of a mountain be 29.4 inches, the attached thermometer 50°, and the heat of the air 45°; at the fame time that at the top of the mountain the barometer is found to ftand at 25.190 inches, the attached thermometer at 46°, and the air-thermometer at 39°½; required the height of the mountain in Englifh feet. Set the numbers down in the following order:

Obſervation at the bottom.

Barometer.	Therm. attached.	Therm. in the air.
In.		
29.400	50°	45°
	46	

Diff. of the two attached thermometers,   4

Obſervation at the top.

	Barom.	Therm. attached.	Therm. in the air.
	In.		
	25.190	46°	39°.½
Correct for the diff. of the two attached therm. *viz.* 4°, }	+ 10		45
			2)84½ (42½ mean heat.
Height of the uppermoſt barometer, reduced to the ſame heat as the lowermoſt, *viz.* 50°, }	25.200		31¼ ſtandard heat,
			11 difference.

	In.	Feet.
Correct for 11°, ſee tab. IV.		
on 4000 feet 106.9	Tabular number, ſee tab. II. corresponding to, } 25.200 = 6225.0	
on 16 — + 5	The ſame, corresponding to 29.400 = 2208.2	
or on 4016 +107.4		
	Approximate height in feet,	4016.8
Correction for 11° of heat on 4016 feet, add, —		107.4
Correct height of the mountain — — —		4124.2

Now the difference of the attached thermometer 50°
and 46° is = 4°; and againſt this number, in table I. in
the column for 25 inches (being the height of the barometer in this caſe) I find 10, which added to 25.190, as
this barometer was the coldeſt, gives 25.200 inches for

I   the

the height of the uppermoft barometer reduced to the fame heat as the lowermoft: and in table II. oppofite to 25 200 inches and 29.400 inches, I find refpectively 6225.0 and 2208.2; their difference 4016.8 is the approximate height in feet. The degrees on the thermometer in the open air, $39°\frac{1}{2}$ and $45°$ being then added together, and afterwards divided by 2, give for the mean temperature of thefe obfervations $42°\frac{1}{4}$, or $11°$ above the ftandard temperature, $31°\frac{1}{4}$: and laftly, the correction for $11°$, in table IV. on 4000 feet I find $= 106.9$, and on 16 feet $= 0.5$; that is, 107.4 feet equal the whole correction, which added to 4016.8 gives 4124.2 feet for the correct height of the mountain.

EXAMPLE II.

Suppofe the height of the barometer at the top of a rock had been obferved at 24.178, the attached thermometer at $57°.2$, the air-thermometer at $56°$; the barometer below at 28.1318 inches, the attached thermometer $61°.8$, the detached one $63°.9$; what is the height of the rock?

Obfer-

Obſervation at the bottom.

Barometer.	Therm. attached.	Therm. detached.
In. 28.1318	61°.8 57.2	63°.9

Difference of the two attached thermometers,   4.6

Obſervation at the top.

	Barom.	Therm. attached.	Therm. detached.
	In. 24.1780	57°.2	56.0 63.9

Correct for the diff. of the two attached therm. *viz.* 4° 6,   } +0112

2)119.9(59.95 mean heat.
          31.24 ſtandard temp.
          ——
          28.71 difference.

Height of the uppermoſt barom. reduced to the ſame heat as the lowermoſt, namely 61°.8,   } 24.1892

		Feet.	Diff.
Tabular number, correſponding to, }	24.1000	7388.0	107.9
The ſame, ſee tab. III.	800 90 2	86.0 9.7 .2 }−95.9	
	24.1892	7292.1	
Tabular number, correſponding to, }	28.1000	3386.6	92.6
The ſame, ſee tab. III.	300 10 8	28.0 0.9 0.7 }−29.6	

Correct for 28°.7, ſee tab. IV.

28° on	{	3000 =	204.1
	{	900 =	61.2
	{	35 =	2.4
0.7 on	{	3000 =	5.1
	{	900 =	1.5
	{	35 =	0.0
28.7 on		3935	274 3

28.1318   3357 0

And 3357.0 feet taken from   —   7292.1

Leaves the approximate height in feet,   3935.1
Correction for 28°7 of heat on 3935 ft.   +274.3

Correct height of this mountain,   4209.4

This

This laſt obſervation was actually made, and the height geometrically was determined to be 4211.3 feet, not quite two feet different. In this example it will be obſerved, that as the height of the barometer is ſet down to four places of decimals; the tabular numbers, anſwering to every tenth only, are corrected by means of table III. of proportional parts, for the remaining decimals 8, 9, and 2, in one place; and 3, 1, 8, in the other; and their ſum is ſubducted from the numbers found in table II. And laſtly, that in finding the correction for 28°.7 of heat, the fraction $\frac{7}{10}$ is conſidered as ſo many units, and another decimal is ſtruck off; thus the correction on 3000 feet for 7° is 51; but for $\frac{7}{10}$ it becomes 5.1, and ſo of the reſt.

EXAMPLE III.

In the upper gallery of the dome of St. Peter's church at Rome, and 50 feet below the top of the croſs, I obſerved the barometer, from a mean of ſeveral obſervations, 29.5218; the thermometer attached being at 56°.6, and the detached one at 57°; at the ſame time that another, placed on the banks of the Tyber one foot above the ſurface of the water, ſtood at 30.0168, the attached thermometer at 60°.6, and the detached one at 60°.2; what was the total height of this building above the level of the river?

Obſer-

Observation below, at one foot above the Tyber.

Barometer.	Therm. attached.	Therm. detached.
In.		
30.0168	6̊0.6	60°.2
	56.6	

Difference of the two attached thermometers, 4.0

Observation above, in the gallery of St. Peter's church.

		29.5218	56.6	57.0
Correct for the diff. of the two attached therm.	}	+ 120		60.2

2)117.2(58.60 mean heat.
　　　 31.24 standard temp.

| Height of the uppermost barom. reduced to the heat of the lowermost *viz.* 60.5, — | } | 29.5338 | | 27.36 difference. |

			Feet.	Diff.
Tabular numbers corresponding to, —	}	29.5000	2119.7	88.2
		300	26.4	
		30	2.6 } —29.7	
		8	7	
		29.5338	2090.0	
Tabular numbers corresponding to, —	}	30.0000	1681.7	86.7
		100	8.7	
		60	5.2 } —14.6	
		8	7	
		30.0168	1667.1	
			2090.0	

Correction for 27°.4		
27° on {	400 = 26 2	
	22 = 1.4	
0.4 on	400 = .4	
27.4 on	422 = 28.0	

Approximate height,	422.9
Correction for 27°4 of heat on 422 feet,	+ 28.0
Difference of height of the barometers,	450.9
Lowest barom. stood 1 foot above the river,	+ 1.0
Top of the cross above the gallery was,	+ 50.0
Total height of the top of the cross above the river Tyber, — }	501.9
The same measured the same day geometrically was, — — }	502.2

When

When the difference of the heights of the quickfilver in the two barometers happens not to exceed $\frac{1}{10}$ or even $\frac{2}{10}$ of an inch (and this will frequently be the cafe in levelling flat countries, or meafuring fmall heights) in fuch circumftances the moft convenient way of reducing the obfervations will be by means of the column of differences only; thofe numbers expreffing the length of a column of the atmofphere which correfponds to $\frac{1}{10}$ of an inch of quickfilver, at any affigned height of the barometer.

E X A M P L E IV.

Suppofe the following obfervations had been made at the top and bottom of any eminence; *viz.* at the top, barometer 29.985 inches, attached thermometer 70°5, detached thermometer 76°; and below, barometer at 30.082, attached thermometer 71°, and the detached one 68°; what was the height of the eminence?

Obfervation below.

Barometer.	Therm. attached.	Therm. detached.
In. 30.0820	$\overset{\circ}{71.0}$ 70.5	$\overset{\circ}{68.0}$

Difference of the two attached therm. 0.5

Obfervation at the top.

	Barometer.	Therm. attached.	Therm. detached.
	In. 29.9850	$\overset{\circ}{70.5}$	$\overset{\circ}{76.0}$
Corre&t for 0°.5 of heat,	+.0015		68.0
Take — —	29.9865		2)144.0(72.0 mean heat.
From — —	30.0820		31.2 ftandard temp.
Remains the difference or fall of quickfilver in the barometer, — }	0.0955		+40.8 difference.

The difference for $\frac{1}{10}$ at 30 inches = 86.7 feet.

	Feet.		Correction for 41°.
Therefore, for 0900 — —	78.0		Feet. Ft.
0050 —	+3		41° on { 80. = 8.0
0005 — —.	0.4.		2.7 = .3
			41° on . 82.7 = 8.3

Therefore, 0955 inch of quickfilver, —	82.7 the approximate height.
Correction for 41° on 82.7 feet,	+8.3
Gives — — — —	91.0 = the true height.

Now this was the height of the Tarpeian rock, or the weft-end of the Capitol-hill in Rome, above the convent of St. Clare, in the *Strada dei fpecchi.*

The preceding rules for determining heights above the furface of the earth will, I prefume, anfwer equally well for meafuring depths below it.

TABLE

TABLE I. For the expansion of quicksilver by heat,
see p. 574.

Degr. of the Therm.	Height of the barometer in inches.												
	20	21	22	23	24	25	26	27	28	29	30	31	32
1	2.0	2.1	2.2	2.3	2.4	2.5	2.6	2.7	2.8	2.9	3.0	3.1	3.2
2	4.1	4.3	4.5	4.7	4.9	5.1	5.3	5.5	5.7	5.9	6.1	6.3	6.5
3	6.1	6.4	6.7	7.0	7.3	7.6	7.9	8.2	8.5	8.8	9.1	9.4	9.7
4	8.1	8.5	8.9	9.3	9.7	10.1	10.5	11.0	11.4	11.8	12.2	12.6	13.0
5	10.1	10.6	11.1	11.6	12.1	12.7	13.2	13.7	14.2	14.7	15.2	15.7	16.2
6	12.2	12.8	13.4	14.0	14.6	15.2	15.8	16.4	17.0	17.6	18.2	18.8	19.5
7	14.2	14.9	15.6	16.3	17.0	17.7	18.4	19.2	19.8	20.6	21.3	22.0	22.7
8	16.2	17.0	17.8	18.6	19.4	20.2	21.0	21.9	22.7	23.5	24.3	25.2	25.9
9	18.2	19.2	20.1	21.0	21.9	22.8	23.7	24.6	25.6	26.5	27.4	28.3	29.2
10	20.3	21.3	22.3	23.3	24.3	25.3	26.3	27.4	28.4	29.4	30.4	31.4	32.4
11	22.3	23.4	24.5	25.6	26.7	27.8	28.9	30.1	31.2	32.3	33.4	34.5	35.6
12	24.3	25.6	26.8	28.0	29.2	30.4	31.6	32.9	34.1	35.3	36.5	37.6	38.9
13	26.3	27.7	29.0	30.3	31.6	32.9	34.2	35.6	36.9	38.2	39.5	40.8	42.1
14	28.4	29.8	31.2	32.6	34.0	35.4	36.8	38.4	39.8	41.2	42.6	43.9	45.4
15	30.4	31.9	33.4	34.9	36.4	37.9	39.4	41.1	42.6	44.1	45.6	47.1	48.6
16	32.4	34.1	35.6	37.2	38.8	40.5	42.0	43.8	45.4	47.0	48.6	50.3	51.8
17	34.5	36.2	37.9	39.6	41.3	43.0	44.7	46.6	48.3	50.0	51.7	53.4	55.1
18	36.5	38.3	40.1	41.9	43.7	45.5	47.3	49.3	51.1	52.9	54.7	56.5	58.3
19	38.5	40.5	42.3	44.2	46.1	48.1	49.9	52.1	54.0	55.9	57.8	59.7	61.6
20	40.6	42.6	44.6	46.6	48.6	50.6	52.6	54.8	56.8	58.8	60.8	62.8	64.9
21	42.6	44.7	46.8	48.9	51.0	53.2	55.2	57.5	59.6	61.7	63.8	65.9	68.1
22	44.6	46.9	49.1	51.3	53.5	55.7	57.9	60.3	62.5	64.7	66.9	69.0	71.4
23	46.6	49.0	51.3	53.6	55.9	58.2	60.5	63.0	65.3	67.6	69.9	72.2	74.6
24	48.6	51.1	53.5	55.9	58.3	60.8	63.1	65.8	68.2	70.6	73.0	75.4	77.8
25	50.7	53.2	55.8	58.2	60.7	63.2	65.7	68.5	71.0	73.5	76.0	78.5	81.1
26	52.7	55.4	58.0	60.5	63.1	65.8	68.3	71.2	73.8	76.4	79.0	81.6	84.3
27	54.7	57.5	60.3	62.9	65.6	68.3	71.0	74.0	76.7	79.4	82.1	84.8	87.5
28	56.8	59.6	62.5	65.2	68.0	70.8	73.6	76.7	79.5	82.3	85.1	87.9	90.7
29	58.8	61.8	64.7	67.5	70.4	73.3	76.2	79.5	82.4	85.3	88.2	91.1	94.1
30	60.8	63.9	66.9	69.9	72.8	75.9	78.9	82.2	85.2	88.2	91.2	94.4	97.3
31	62.8	66.0	69.1	72.1	75.2	78.4	81.5	84.9	88.0	91.1	94.2	97.4	100.5
32	64.8	68.2	71.4	74.6	77.7	81.0	84.2	87.7	90.9	94.1	97.3	100.5	103.8
33	66.9	70.3	73.6	76.9	80.1	83.5	86.8	90.4	93.7	97.0	100.3	103.6	107.0
34	68.9	72.4	75.8	79.2	82.5	86.1	89.4	93.2	96.6	100.0	103.4	106.7	110.3
35	70.9	74.5	78.0	81.5	84.0	88.6	92.0	95.9	99.4	102.9	106.4	109.9	113.5
36	73.0	76.7	80.2	83.8	86.4	91.1	94.6	98.6	102.2	105.8	109.4	113.1	116.8
37	75.0	78.8	82.5	86.2	88.9	93.6	97.3	101.4	105.1	108.8	112.5	116.2	120.0
38	77.0	80.9	84.7	88.5	91.3	96.2	99.9	104.1	107.0	111.7	115.5	119.3	123.2
39	79.0	83.1	86.9	90.8	93.7	98.7	102.9	106.9	110.8	114.7	118.6	122.5	126.5
40	81.1	85.2	89.2	93.2	97.2	101.2	105.2	109.6	113.6	117.6	121.6	125.6	129.7

TABLE II[(f)]. Giving the approximate height in Englifh feet, adapted to the temperature 31°24 of FAHREN-HEIT's thermometer.

Height of the Barom.	Height.	Diff.	Height of the Barom.	Height.	Diff.	Height of the Barom.	Height.	Diff.
Inch.	Feet.		Inch.	Feet.		Inch.	Feet.	
1.—	90309.0	18062	16.10	19570.4	173.1	16.60	17102.5	157.5
2.—	72247.2	10565	20	19398.4	172.0	70	16946.0	156 5
3.—	61681.8	7496	30	19227.5	170.9	80	16790.4	155.6
4.—	54185 4	5814	40	19057.7	169.8	90	16635.8	154.6
5.—	48370 8	4761	50	18889.1	168.6	17 00	16482.1	153 7
6.—	43619.9	4017	60	18721.5	167 6	10	16329.2	152.9
7.—	39603.1	3480	70	18555.0	166.5	20	16177.3	151.9
8.—	36123.6	3069	80	18389.6	165.4	30	16026.2	151.1
9.—	33054.4	2745	90	18225.2	164.4	40	15876.0	150.2
10.—	30309.0	2484	16.00	18061.8	163 4	50	15726.7	149.3
11.—	27825.4	2267	10	17899.4	162.4	60	15578.2	148.5
12.—	25558.1	2086	20	17738.1	161.3	70	15430.6	147.6
13.—	23472.4	1931	30	17577.7	160.4	80	15283.8	146.8
14.—	21541.3	1798	40	17418.4	159.3	90	15137.8	146.0
15.00	19743.5		50	17260.0	158.4	18.00	14992 6	145.2

(f) This table bears fome analogy to the tables of logiftical logarithms, being nothing more than the differences of the logarithms. of the height of the barometer from the logarithm of 32 inches multiplied by fix. I have chofen the logarithm of 32 for my term of comparifon, that being the greateft probable height that the barometer will ever be feen as, even at the bottom of the deepeft mines. Had I taken the mean height of the quickfilver at the level of the fea, it is true the numbers in the table would have more truly reprefented the heights in the atmofphere, correfponding to the given height of the quick-filver; but then, in computing fmall depths or heights from the furface of the fea, we fhould have been obliged fometimes to have changed the figns in the operation, which appeared to me lefs convenient. The mean height of the barometer at the level of the fea, from 132 obfervations in Italy and in England, is 30.04 inches, the heat of the barometer being 55°, and the air 62°; fo that the term of comparifon in this table, *viz.* 32 inches, correfponds to an imaginary point within the earth at 1647 feet below the furface of the fea.

TABLE

TABLE II. continued.

Height of the Barom.	Height.	Diff.	Height of the Barom.	Height.	Diff.	Height of the Barom.	Height.	Diff.
Inch.	Feet.		Inch.	Feet.		Inch.	Feet.	
18,10	14848.3	144.3	22.00	9763.6	118,8	25.90	5511.0	100.8
20	14704.7	143.6	10	9645.5	118,1	26.00	5410.4	100.6
30	14561.9	142.8	20	9527.8	117.7	10	5310.6	99.8
40	14419.9	142.0	30	9410.7	117,1	20	5210.9	99.7
50	14278.7	141.2	40	9294.1	116.6	30	5111.6	99.3
60	14138.2	140.5	50	9178.1	116:0	40	5012.8	98.8
70	13998.5	139.7	60	9062.5	115.6	50	4914.2	98.6
80	13859.5	139.0	70	8947.4	115,1	60	4816.1	98 1
90	13721.3	138.2	80	8832 9	114.5	70	4718.3	97 8
19.00	13583.8	137.5	90	8718.9	114.0	80	4620.9	97 4
10	13447.0	136.8	23.00	8605.3	113.6	90	4523.9	97.0
20	13310.9	136.1	10	8492.3	113.0	90	4427.2	96.7
30	13175.6	135.3	20	8379.7	112.6	27.00	4330.8	96.4
40	13041.1	134.5	30	8267.6	112.1	10	4234.9	95.9
50	12906.9	134.2	40	8156.0	111.6	20	4139.2	95.7
60	12773.6	133.3	50	8044.9	111.1	30	4044.0	95.2
70	12641.0	132.6	60	7934.3	110.6	40	3949.0	95.0
80	12509.1	131.9	70	7824.1	110.2	50	3854.5	94.5
90	12337.0	131.3	80	7714.4	109.7	60	3760.2	94.3
20.00	12247.2	130.6	90	7605.1	109.3	70	3666.3	93 9
10	12117.2	130.0	24.00	7496.3	108.8	80	3572.7	93 6
20	11987.9	129.3	10	7388.0	108.3	90	3479.5	93.2
30	11859.2	128.7	20	7280.1	107.9	28.00	3386.6	92 9
40	11731.2	128 0	30	7172.6	107.5	10	3294.0	92.6
50	11603.8	127.4	40	7065.6	107.0	20	3201.8	92 2
60	11477.0	126.8	50	6959.0	106.6	30	3109.9	91.9
70	11350.0	126.2	60	6852.9	106.1	40	3018.3	91.6
80	11225.2	125.6	70	6747.2	105.7	50	2927.0	91.3
90	11100.2	125 0	80	6641.9	105.3	60	2836.1	90.9
21.00	10975.8	124.4	90	6537.0	104.9	70	2745.4	90.7
10	10852.1	123.7	25.00	6432.6	104.4	80	2655.1	90.3
20	10728.8	123.3	10	6328.6	104.0	90	2565.1	90.0
30	10606.2	122.6	20	6225.0	103.6	29.00	2475.4	89.7
40	10484.2	122.0	30	6121.8	103.2	10	2386.0	89.4
50	10362.7	121.5	40	6019.0	102.8	20	2296.9	89.1
60	10241.8	120.9	50	5916.6	102.4	30	2208.2	88.5
70	10121.4	120.4	60	5814.6	102.0	40	2119.7	88 2
80	10001.6	119 8	70	5713.0	101.6	50	2031.5	87.9
90	9882.4	119.2	80	5611.8	101.2	60	1943.6	

TABLE

TABLE II. continued.

Height of the Barom.	Height.	Diff.	Height of the Barom.	Height.	Diff.	Height of the Barom.	Height.	Diff.
Inch.	Feet.		Inch.	Feet.		Inch.	Feet.	
29.80	1856.0	87.6	30.60	1165.7	85.3	31.40	493.2	83.1
90	1768.7	87.3	70	1080.7	85.0	50	410.4	82.8
30.00	1681.7	87.0	80	996.0	84.7	60	327.8	82.6
10	1595.0	86.7	90	911.5	84.5	70	245.4	82.4
20	1508.6	86.4	31.00	827.3	84.2	80	163.4	82.0
30	1422.4	86.2	10	743.4	83.9	90	81.6	81.0
40	1236.6	85.8	20	659.7	83.7	32.00	00.0	81.6
50	1251.0	85.6	30	576.3	83.4			

TABLE

T A B L E III. Of proportional parts.

Diff.	1	2	3	4	5	6	7	8	9	Diff.	1	2	3	4	5	6	7	8	9
81	8	16	24	32	40	49	57	65	73	106	11	21	32	42	53	64	74	85	95
82	8	16	25	33	41	49	57	66	74	107	11	21	32	43	53	64	75	86	96
83	8	17	25	33	41	50	58	66	75	108	11	22	32	43	54	65	76	86	97
84	8	17	25	34	42	50	59	67	76	109	11	22	33	44	54	65	76	87	98
85	8	17	25	34	42	51	59	68	76	110	11	22	33	44	55	66	77	88	99
86	9	17	26	34	43	52	60	69	77	111	11	22	33	44	55	67	78	89	100
87	9	17	26	35	43	52	61	70	78	112	11	22	34	45	56	67	78	90	101
88	9	18	26	35	44	53	62	70	79	113	11	23	34	45	56	68	79	90	102
89	9	18	27	36	44	53	62	71	80	114	11	23	34	46	57	68	80	91	103
90	9	18	27	36	45	54	63	72	81	115	11	23	34	46	57	69	80	92	103
91	9	18	27	36	45	55	64	73	82	116	12	23	35	46	58	70	81	93	104
92	9	18	28	37	46	55	64	74	83	117	12	23	35	47	58	70	82	94	105
93	9	19	28	37	46	56	65	74	84	118	12	24	35	47	59	71	83	94	106
94	9	19	28	38	47	56	66	75	85	119	12	24	36	48	59	71	83	95	107
95	9	19	28	38	47	57	66	76	85	120	12	24	36	48	60	72	84	96	108
96	10	19	29	38	48	58	67	77	86	121	12	24	36	48	60	73	85	97	109
97	10	19	29	39	48	58	68	78	87	122	12	24	37	49	61	73	85	98	110
98	10	20	29	39	49	59	69	78	88	123	12	25	37	49	61	74	86	98	111
99	10	20	30	40	49	59	69	79	89	124	12	25	37	50	62	74	87	99	112
100	10	20	30	40	50	60	70	80	90	125	12	25	37	50	62	75	87	100	112
101	10	20	30	40	50	61	71	81	91	126	13	25	38	50	63	76	88	101	113
102	10	20	31	41	51	61	71	82	92	127	13	25	38	51	63	76	89	102	114
103	10	21	31	41	51	62	72	82	93	128	13	26	38	51	64	77	90	102	115
104	10	21	31	42	52	62	73	83	94	129	13	26	39	52	64	77	90	103	116
105	10	21	31	42	52	63	73	84	94	130	13	26	39	52	65	78	91	104	117

T A B L E

TABLE IV.   For the expanfion of the air, or correction of the uppermoft height, fee p. 576.

Deg.	Approximate height in feet.								
°	1000.	2000.	3000.	4000.	5000.	6000.	7000.	8000.	9000.
1	2.4	4.9	7.3	9.7	12.1	14 6	17.0	19.4	21.9
2	4.9	9.7	14.6	19 4	24.3	29.2	34 0	38.9	43.7
3	7.3	14.6	21.9	29.2	36.4	43.7	51.0	58.3	65.6
4	9.7	19.4	29.2	38.9	48.6	58.3	68.0	77.8	87.5
5	12.1	24.3	36.4	48.6	60.7	72.9	85.0	97.2	109.3
6	14.6	29.2	43.7	58.3	72.8	87.5	102.0	116.6	131.2
7	17.0	34.0	51.0	68.0	85.0	102.1	119.0	136.1	153.0
8	19.4	38.9	58.3	77.8	97.1	116.6	136.0	155.5	174.9
9	21.9	43.7	65.6	87.5	109.3	131.2	153.0	175.0	196.8
10	24.3	48.6	72.9	97.2	121.5	145 8	170.1	194.4	218.7
11	26.7	53.5	80.2	106.9	133 6	160.4	187.1	213.8	240.6
12	29.2	58.3	87.5	116.6	145.8	175.0	204.1	233.3	262.4
13	31.6	63.2	94.8	126.4	157.9	189.5	221.1	252.7	284.3
14	34.0	68.0	102.1	136.1	170.1	204.1	238.1	272.2	306.2
15	36.4	72.9	109.3	145.8	182.2	218.7	255.1	291.6	328.0
16	38.8	77.8	116.6	155.5	194.3	233.3	272.1	311.0	349.9
17	41.3	82.6	123.9	165.2	206.5	247.9	289.1	330.5	371.7
18	43.7	87.5	131.2	175.0	218.6	262.4	306.1	349.9	393.6
19	46.1	92.3	138.5	184.7	230.8	277.0	323.1	369.4	415.5
20	48.6	97.2	145.8	194.4	243.0	291.6	340.2	388.8	437.4
21	51.0	102.1	153.1	204.1	255.1	306.2	357.2	408.2	459.3
22	53.5	106 9	160.4	213.8	267.3	320.8	374 2	427.7	481.1
23	55.9	111 8	167.7	223.6	279.4	335.3	391.2	447 1	503.0
24	58.3	116.6	175.0	233.3	291 6	349.9	408.2	466.6	524.9
25	60.7	121.5	182.2	243.0	303.7	364 5	425 2	486 0	546.7

I   TABLE

TABLE IV. continued.

Deg.	Approximate height in feet.								
o	1000	2000.	3000.	4000	5000.	6000.	7000.	8000.	9000.
26	63.1	126.4	189.5	252.7	315.8	379.1	442.2	505.4	568.6
27	65.6	131.2	196.8	262.4	328.0	393.7	459 2	524.9	590.4
28	68.0	136.1	204 1	272.2	340.1	408.2	476.2	544.3	612 3
29	70.4	140.9	211.4	281.9	352 3	422.8	493.2	563.8	634.2
30	72.9	145.8	218.7	291 6	364.5	437.4	510.3	583.2	656.1
31	75.3	150.7	226.0	301.3	376.6	452.0	527.3	602.6	678.0
32	77.8	155 5	233.3	311.0	388.8	466.6	544.3	622.1	699.8
33	80.2	160.4	240.6	320 8	400 9	480 1	561.3	641.5	721.7
34	82.6	165.2	247.9	330.5	413 1	495.7	578.3	661 0	743.6
35	85 0	170.1	255.1	340.2	425.2	510.2	595.3	680.4	765.4
36	87.4	175.0	262.4	349.9	437.3	524.8	612.3	699.8	787.3
37	89.9	179.8	269.7	359.6	449.5	539.4	629.3	719.3	809.1
38	92.3	184.7	277.0	369 4	461.6	553.9	646.3	738.7	831.0
39	94.7	189.5	284.3	379.1	473.8	568.5	663.3	758.2	852.9
40	97.2	194.4	291.6	388.8	486.0	583.2	680.4	777.6	874.8
41	99 6	199.3	298 9	398 5	498.1	597.8	697.4	797.0	896.7
42	102.1	204.1	306.2	408.2	510.3	612 4	714.4	816.5	918.5
43	104.5	209.0	313 5	418.0	522.4	626.9	731.4	835.9	940.4
44	105.9	213.8	320 8	427.7	534.6	641.5	748.4	855 4	962 3
45	109.3	218.7	328 0	437.4	546.7	656.1	765 4	874.8	984.1
46	111.7	223 6	335.3	447.1	558.8	670.7	782 4	894.2	1006.0
47	114.2	228.4	342 6	456 8	571.0	685.3	799 4	913.7	1027.8
48	116.6	233.3	349.9	466.6	583.1	699.8	816 4	933.1	1049.7
49	119.0	238.1	357.2	476.3	595.3	714 4	833.4	952.6	1071.6
50	121.5	243.0	364.5	486.0	607.5	729.0	850.5	972 0	1093.5

Table

Table of heights taken by the barometer, &c.

	+ or — the Lake of Geneva. Feet.	Above the Mediterranean. Feet.
The Lake of Geneva, from 18 observations, —	0	1230 (g)
Greatest depth of the Lake, — —	— 393	
Cluse, at the Croix Blanche, first-floor, (h) 2,	+ 351	1581
Chamouny, ground-floor of the inn near the foet of Mont Blanc, 4 — — —	+ 2137	3367
The Montanvert, at the Chateâu, 1 —	+ 5001	6231
The source of the river Arvéron, at the bottom of the Vallée de Glace, 1 — — —	+ 2426	3656
Salenche, at the inn, second-floor, 1 —	+ 664	1948
La Bonne-Ville, a la Ville de Geneve, second-floor, 1	+ 245	1475
Chatlaino, country house near Geneva, ground-floor, 0	+ 178	
The ball on the highest, or south-west, tower of St. Peter's church in Geneva, 0 — —	+ 449	
St. Joire, in a field at the foot of the Mole, 0 —	+ 671	1901
Summit of the Mole, — — — —	+ 4883	6113
Pitton, highest point of Mont Saleve, 0 —	3284	4514
The Dole, highest summit of Mont Jura, 0 —	+ 4293	5523
The Buet, 0 — — — —	+ 8894	10124
Aiguille d'Argentiére, 0 — — —	+ 12172	13402
Mont Blanc, 0 — — —	+ 14432	15662
Frangy, at the inn, first-floor, below the Lake,	— 166	
Aix, a la Ville de Geneve, first-floor, below the Lake,	— 378	
Chambery, au St. Jean Baptiste, first-floor, below the Lake,	— 352	
Aiguebelle, at the inn, first-floor, below the Lake,	— 190	
La Chambre, at the inn, first-floor, above the Lake,	+ 337	
St. Michael, at the inn, first-floor, —	+ 1113	2343
Modane, at the inn, first-floor, — —	+ 2220	3450

(g) More correctly 1228 feet, but I have taken it at 1230 in round numbers.

(h) The figures at the end of some of the names shew the number of observations that were made; and the letter o indicates such observations to have been geometrical.

Table

Table of heights, &c. continued.

	+ or — the Lake of Geneva.	Above the Mediterranean.
	Feet.	Feet.
Lannebourg, the foot of Mont Cenis, at the inn, first-floor,	+ 3178	4408
Mont Cenis, at the Post , — — —	+ 5031	6261
————— at the Grande Croix, — —	+ 4793	6023
Novalese, the foot of Mont Cenis on the side of Italy, at the inn, first-floor, — —	+ 1511	2741
Boucholin, on the first-floor, — —	+ 213	
St. Ambroise, on the first-floor, below the Lake,	— 40	
Turin, à l'Hôtel d'Angleterre, second-floor, 4	— 289	941
Felissano, near Alessandria, first-floor, 1 —	— 671	
Piacenza, St. Marco, first-floor, 1 — —	— 967	263
Parma, au Paon, first-floor, 3 — —	— 923	307
Bologna, au Pelerin, first-floor, 3 —	— 831	399
Loiano, a little village on the Appenines, between Bologna and Florence, — —		2591
The mountain Raticosa, the highest point of the Appenines the road passes over, 1¼ miles beyond File- caije in going to Covigliaje,	+ 1671	2901
Florence, nel Corso dei Tintori, 50 feet above the Arno, which was 18 feet below the wall of the quay, 3	— 990	+ 240
Pisa, aux Trois Demoiselles, second-floor, 4 —	— 1228	+ 541
Leghorn, chez Muston, second-floor, 2 —	— 1244	+ 38
Siena, aux Trois Rois, second-floor, 2 —	— 164	1066
Redicoffani, at the Post, first-floor, above the Lake,	+ 1240	2470
————— the top of the tower of the old fortifica- tion on the summit of the rock, —	1830	3060
Viterbo, aux Trois Rois, first-floor, on the Ciminus of the Ancients, — — —	+ 29	1259
Rome, nel Corso, 61 feet above the Tyber, 7	— 1084	94

(i) The rocks on each side the plain, where the post-house stands, are at least 3000 feet higher than this situation; and it is from the snow on the tops, and through the crevices, that the lake on this plain is formed, which gives rise to the Dora, and may be called one of the sources of the Po.

Table of heights, &c. continued.

	Above the River Tyber. Feet.	Above the Mediterranean. Feet.
The Level of the river Tyber, — —		33
The top of the Janiculum, near the Villa Spada, —	260	
Aventine Hill, near the Priory of Malta, —	117	
In the Forum, near the arch of SEVERUS, where the ground is raised 23½ feet,	34	
Palatine Hill, on the floor of the Imperial palace —	133	
Celian Hill, near the CLAUDIAN aqueduct, —	125	
Bottom of the canal of the CLAUDIAN aqueduct, —	175	
Esquiline Hill, on the floor of St. M. Major's church,	154	
Capitol Hill, on the West-end of the Tarpeian rock,	118	
In the Strada dei Specchi, in the convent of St. Clare,	27	
On the union of the Viminal and Quirinal Hills, in the Carthufian's church, DIOCLES. Baths, —	141	
Pincian Hill, in the garden of the Villa Medici, —	165	
Top of the crofs of St. Peter's church, —	502	
The bafe of the obelisk, in the center of the Periftyle,	31	
The fummit of the mountain Soracte, lying about 20½ geog. miles N. of Rome, G		2272
The fummit of Monte Velino, one of the Appenines, covered with fnow in June, about 46 geog. miles N.W. of Rome, and which is probably the higheft of the Appenines, G — — —		8397

	+ or — the Lake of Geneve.	
Naples, Cafa Ifolata on the Chiaia, 27½ feet above the fea, 5 — — —	— 1197	
Mount Vefuvius, mouth of the Crater from whence the fire iffued in 1776, — —		3938 (1)

Table

(*l*) Sir WILLIAM HAMILTON informed me, that the height of Vefuvius, as taken by Mr. DE SAUSSURE of Geneva in 1772, with only a barometer of Mr. DE LUC's conftruction, and according to his rules, was 3659½ French feet = 3900 Englifh, which agrees pretty well with mine. But the Padre DELLA TORRE pretends to have found the height of Vefuvius in 1752 (fee p. 44. of his

Table of heignts, &c. continued.

	+ or — the Lake of Geneva.	Above the Mediterranean.
Mount Vesuvius, at the base of the cone,	— —	2021
Top of the mountain Somma,	— —	3738
The summit of Mount Ætna,	— —	10954 (l)

The following heights are determined from corresponding observations by Mr. MESSIER at Paris, whose barometer is supposed 108 feet above the sea.

Barberino di Valdensa, between Boggebonri and Tavernelle,	974
Modena, a l'Albergo nuovo, — —	214
Montmelian, at 20 feet above the river, — —	811
Monte Viso, by an observation from Jurin, by means accurate, G — —	9997
Monte Rosa, as measured geometrically by the Father BECCARIA, being the second mountain of all the Alps, — — — —	15084
Pont Beauvoisin, — — — —	705
La tour du Pin, 4 — — —	938
Verpilliére, — — — —	566

his History of this Mountain) = 1677 French feet only, the difference of his barometer at the top and at the level of the sea being no more than 23½ French lines = 2.065 English inches, which was certainly a mistake of little less than 2000 feet in the result. The Abbé NOLLET in 1749 found the fall of the quicksilver 40 lines = 3.55 inches English; and, if this observation is to be depended upon, the summit of this volcano has risen within these 27 years more than 330 feet perpendicular.

(l) I have ventured to compute the height of this celebrated mountain from my own tables, though from an observation of Mr. DE SAUSSURE's in 1773, which that gentleman obligingly communicated to me. It will serve to shew, that this volcano is by no means the highest mountain of the old world; and that Vesuvius, placed upon Mount Ætna, would not be equal to the height of Mont Blanc, which latter I take to be the most elevated point in Europe, Asia, or Africa.

The circumference of the visible horizon on the top of Mount Ætna, allowance being made for refraction, which I estimate at 6', is 1093 English miles.

Table

Table of heights, &c. continued.

	+ or — the Lake of Geneva.	Above the Mediterranean.
Lyons, at the Hôtel Blanc, 50 feet above the Saône,		449
St. Jean le vieux, — — —		695
Cerdon, near the post-house at the foot of the rocks,		854
Nantua, 10 feet above the Lake, — —		1423
Chatillon, at the Logis Neuf, — —		1629
Colonges, — — —		1626
St. Genis, apparently on a level with the foot of Mont Jura,		1501
Geneva, at 100 feet above the Lake, 5 —		1268(*m*)
Mâcon, at the Parc, 24 feet above the Saône —		514
Dijon, à la Cloche, the first-floor, — —		710
Mountain of Maraifelois (*n*), 4¾ miles beyond Viteaux towards Dijon, — — }		1677
Lucy-le-bois, — — — —		645
Auxerre, 50 feet above the river, — —		283
Sens, at the Post, — — —		163
Fontainbleau, at the Grand Cerf, second-floor, —		242

(*m*) From this comparison with Mr. MESSIER's observations at Paris, which makes the Lake of Geneva only 1768 feet above the level of the sea (whereas from 18 observations in Italy, near the shore of the Mediterranean, it appears to be 1228; viz. +60 feet different) I am inclined to believe, that Mr. MESSIER's place of observation is about 50 feet higher than I have supposed it, viz. 160 feet above the sea instead of 108, as deduced from three observations only at Boulogne, Calais, and at Dover. If this be allowed, the same number of feet must be added also to all the other heights that are determined by comparison with Mr. MESSIER's observations. I am, however, by no means sure of this, but leave it to future observers.

(*n*) On one side of this mountain is a little stream called Amancon, that joins the Yonne and the Seine, and thus goes to the Atlantic; while on the other side is found the Ouche, which, uniting with the Saône and the Rhone, runs to the Mediterranean: this part of Burgundy then seems to be one of the highest in France.

Table

Table of heights, &c. continued.

	+ or — the Seine at Paris.	Above the Mediterranean.
Paris, mean height of the Seine, that is, *quand les eaux fe trouvent à 13 pieds 9 pouces fur l'echelle du Pont Royal felon M. DE LA LANDE;*		36¼
Place of my own obfervations in the Ruë Jacob, fecond-floor,	+ 57	
Mr. MESSIER's obfervatory, at the Hôtel de Clugny, firft-floor,	72	
Mr. DE LA LANDE's ditto, at the College Royal, firft-floor,	101	
Place of Monf. le Pere COTTES's obfervations at Mont-morency, 10 miles North of Paris,	333	
Stone-gallery of the Church on Mont Valerien,	473	
Depth of the cave of the Royal Obfervatory at Paris below the pavement,	98¼	
The fame, according to Mr. DE LA LANDE, by actual meafurement,	98	
Height of the north tower of the church of Notre Dame above the floor,	220⅘	
—— by actual meafurement,	218⅘	
Chantilly,		119
Clermont,		329
Amiens, Ruë de Noyon, firft-floor,		147
Abbeville, firft-floor,		79

	Below the mean height of the Seine.	
Boulogne, mean level of the fea, from one obferv. only,	—33.9	
Calais, ditto, from one obfervation,	—38.8	
Dover, ditto, from three obfervations made two years preceding thofe at Calais and Boulogne,	—36.6	
Mean height of the river (o) Thames at London above the mean height of the river Seine from five direct comparifons with Mr. MESSIER,	+ 6.8	
And confequently the Thames at London above the fea,		43
Warwick, mean level of the river Avon,		155
Shuckburgh-houfe, in Warwickfhire,		560

(o) By the mean height of the river Thames is underftood when the water is 15¼ feet below the pavement in the left-hand arcade at Buckingham-ftairs.

Table of the Angles & Sides of the different Triangles.

Place of Observation	Object	Horizontal Angle	Error in the Angle	Distance in English Feet	Corrected Angle with the Horizon	Difference of height in Feet	Lines of Error in Feet	Height above the Lake of Geneva
The end A of the Base A B	Pitton of Saleve and B	58.28.10	6	13286.4	+10.29.14 27	7835.1	3	3294.2
				1760.8	+0.26.49	22.3		
	Pitton & Church St Pierre low was above y Lake	157.13.30	30					
				24486.5	−0.31.35 15	224.3	3	
	Tower of Arthain & B	129.10.1		2172				
End B of the Base	A & Pitton	11 .... 111.52.16	6					
				14041.7	+11.17.41 26	2806.3	3	3287.6
	A & Tower of Arthain	22.10.0						
The Pitton or highest point of Mt Saleve	A & B	9.38.35	6					
	A & Church St Pierre	17.47.0	30					
	The Mole & Church St Pierre was above y Lake	83.21.45	30	78913.7	+1.2.6 1.15	1596.		4879.6
				38593.3	−1.33.22 4	3039.5		3282.6
	Mt Blanc & The Mole	30.16.34	60	206879.	+2.47.57 3.27	11124.	64	14411.7
	Glacier le Buet & Church St Pierre	94.16.52	30	182446.	+1.30.16 2.24	5613.6		8899.2
	Aiguille d'argent & Mole	18.18.45	45	217723.	+2.2.12 3.40	8878.4		12162.0
	Varens & Church St Pierre	44.27						
	N.E. end of Saleve & Church St Pierre	32.34						
	Pitton Spierre was y Lake	26.56.0	30					

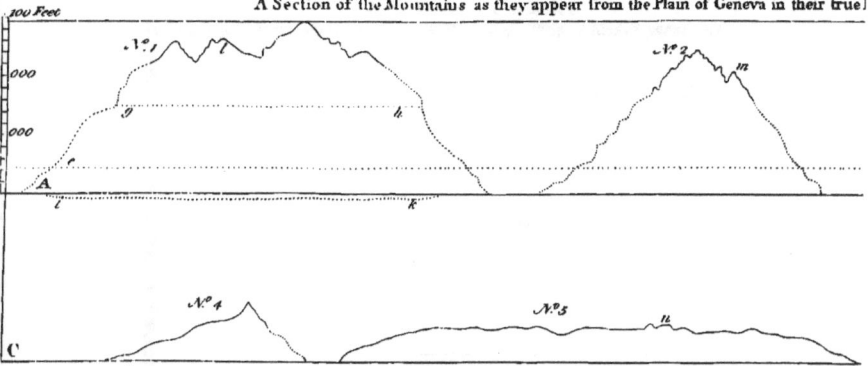

A Section of the Mountains as they appear from the Plain of Geneva in their true]

100 Feet

000

000

A

N.º 1

N.º 2

g

h

c

l

k

m

C

N.º 4

N.º 5

h

AB & C D. represent the Level with the Lake of Geneva. N.º 1. Mont. Blanc. 2. The Aiguille d'Argentiere. 3
5. Mont Saleve. e f. The Level of the Valey of Chamouny, the foot of Mont Blanc. g. h. A line that expresses t
lies constantly the whole Summer. i.k. shews the depth of the Lake (according to M.. M.) proportionaly to the Moun
widest part. l. the point to which 4 Inhabitants of Chamouny relate to have ascended in 1775. m. (in N.º 2) suppos
the Mer de Glace, in the Valley of Chamouny. N. the Pitton of Saleve.

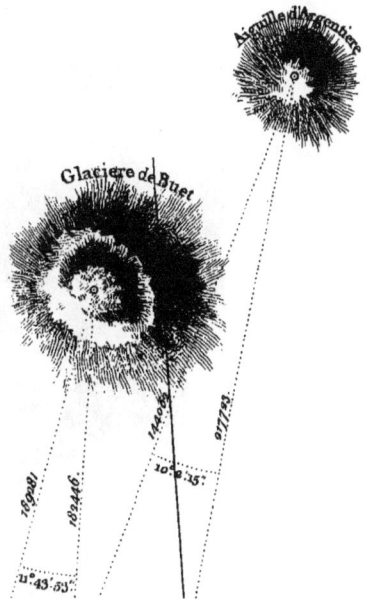

Aiguill d'Argentiere

Glaciere de Buet

16981

18446

14406

17728

10º 9 15

11 43 53

ıeir true Proportions.

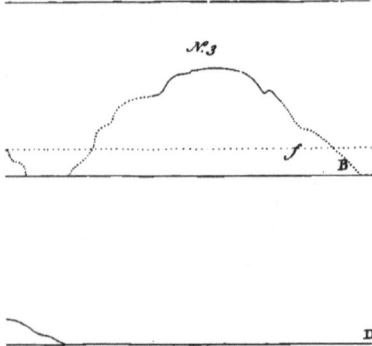

Nº 3

f   B

D

ntière . 3. *The Glaciere de Buet* . 4. *The Mole* ;
hrɽʃas *the Limit above which the Snow*
he *Mountains; being a Section of it in the*
ɟ *supposed to be the Aiguille de Dru, near*

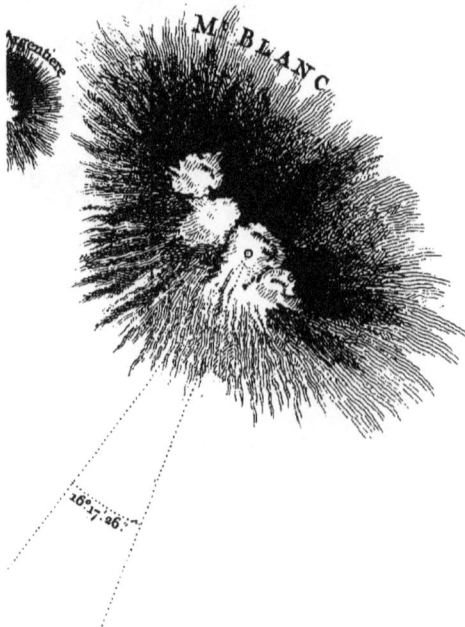

Mt. BLANC

gentière

16°. 17. 26.

wroh S.Pierre at Geneva	Varons & Pitton	106.30	....	....	....	....
	Pitton & Monetier	50.0	....	....	....	....
	Pitton & Little Saleve	54.54	....	....	....	....
	Pitton & Fort la Cluse	61.30	....	....	....	....
	Pitton & S.W. end of Saleve	10.0	....	....	....	....

The Azimuth of the Mole from St Pierre is 66°.9′.27″.S.E.

Note The dotted Figures in the Column entitled Corrected Angle with the Horizon, express the Supposed effect of Refraction which has been made use of in the Computations.

Heights finally reduced.

Saleve	3283.6	
Dole	4292.7	
Mole	4882.8	Mean heights
Buet	8693.7	above the Lake
Argentiere	12171.8	in English Feet
Mont Blanc	14432.5	

The Height of the Lake is reckoned 3 feet 9 Inches below the Summit of the North Pierre du Niton; 3.9 below the South one; & lastly 2 fg. 1 below the Base of the S.W. Tower of St Peters Church.

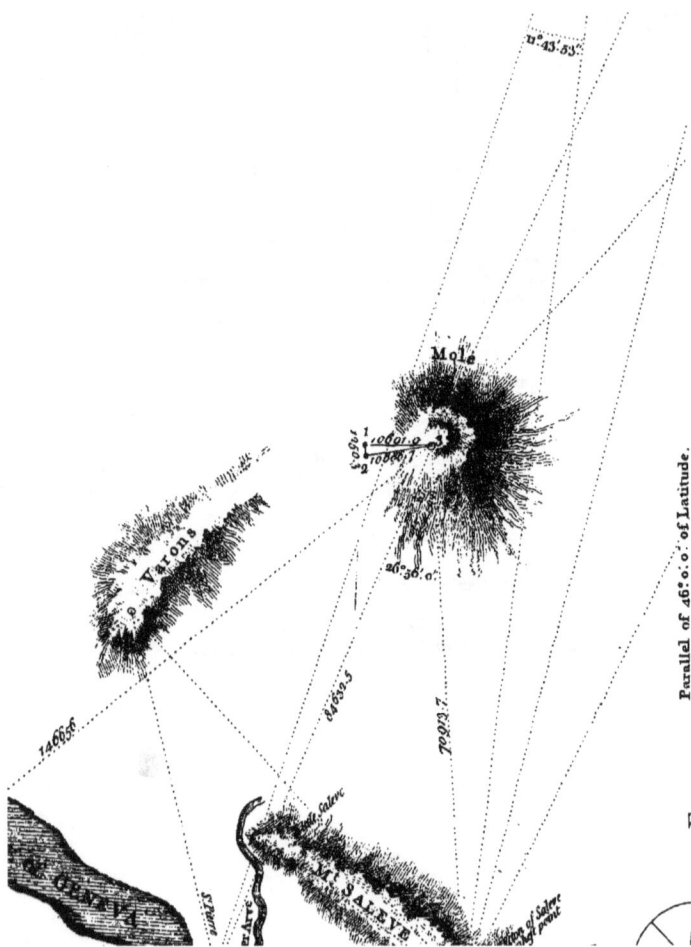

Meridian of Geneva

A CHART

Shewing the

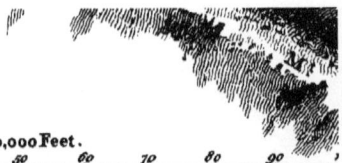

Scale of 100,000 Feet.

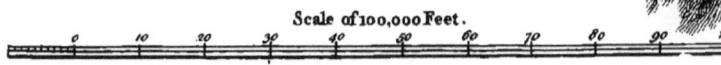

English Miles.

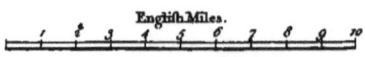

Minutes of a Degree of Latitude.

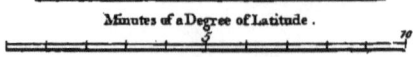

Minutes of Longitude in Lat 46.0.

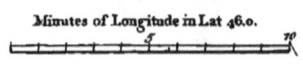

Shewing the

SITUATION, DISTANCES and HEIGHTS

of some of the most remarkable

MOUNTAINS,

that are seen from the borders of the

LAKE OF GENEVA.

Surveyed in Aug.t & Sept.r 1775.

G. S.

Basire Sculp.

E R R A T A.

Page Line

58, 9. *for* commmunicate, *read* communicate,

85, 15. *for* XLVIII, *read* LIV.

128, 4. from the bottom, for " and not all" *read* " and not at all"

131, 16 and 17. *for* (as the millers term it when no Iron is concerned) *read* (as the millers term it) where no iron is concerned

162, 6. *for* Satellites, *read* Satellite.

165, 9. *for* ineptas, *read* ineptas

258, 3. from the bottom, *for* but, *read* long

258, 2. from the botttom, *for* long, *read* but

354, 2. *for* the year 1775, *read* the year 1776.

475, 13. *for* credulicity, *read* credulity

518, 7. from the bottom, *for* $\dfrac{1}{2000}$ *read* $\dfrac{1}{20,000}$.

519, 7. *for* 233°, 54', 15" *read* 233°, 53'. 15"

520, 4. *insert* ∠ c by 4th observation $=9°$, 59' 0"$—9°$, 38', 15'''

521, 2. *for* mountains, *read* mountain.

522, 2. *for* correct for the signal 59", *read* 54"

530, 5. *for* 27,7025; *read* 25,7025

541, 4. *for* above at C. *read* above at B.

545, 11. *for* correct height in fathom 686,619, *read* 685,619

546, 8. *for* difference of Log. 654,157. *read* 654,109.

547, 11. *for* (in p. 556), *read* (in p. 532).

556, 17. *for* two, *read* too

560, 1. *for* feet, *read* grains

  13. *for* 13358,5, *read* 13558,5.

562, 12. *for* barometer, *read* manometer

568, 5. from the bottom, *for* $\overline{T-S} \times \overline{E-t}-a = S-x$, *read* $\dfrac{\overline{T-S} \times \overline{E-t}-a}{E} = S-n.$

569, 19. *dele* the semicolon after quantity, and insert it after instance

578, 5. from the bottom, *for* the attached Therm, *read* the two attached Therm.

585, 2. *read*, see p. 574 and 567

  in the column for 25 inches, and against 21 *for* 53,2, *read* 53,1

586, 3. *add*, see p. 568 and 569

  In the 4th col. of the table at the top, *for* 16,10, *read* 15,10.

A CHART
Shewing the
SITUATION DISTANCES and HEIGHTS
of Some of the most remarkable
MOUNTAINS,
that are seen from the borders of the
LAKE of GENEVA

www.ingramcontent.com/pod-product-compliance
Lightning Source LLC
Chambersburg PA
CBHW021948190326
41519CB00009B/1179